D0646902

THE STORY OF NEUROSCIENCE

Unlocking the mysteries of the brain and consciousness

THE STORY OF NEUROSCIENCE

Anne Rooney

This edition published in 2017 by Arcturus Publishing Limited
26/27 Bickels Yard, 151–153 Bermondsey Street,
London SE1 3HA

ISBN: 978-1-78428-536-4
AD005417UK

Printed in China

CONTENTS

Introduction: MIND AND BODY

'If the human brain were so simple that we could understand it, we would be so simple that we couldn't.'

Emerson Pugh, philosopher, 1938

As you read this book, your brain is working hard. It's not just processing the information you're reading, it's taking the input from your eyes and turning it into information and forming memories. It's getting your fingers to turn the pages and moving your eyes along the lines of text. If someone later asks what you've read, it will enable you to understand the question and formulate an answer. And, as always, your brain and nervous system are controlling your heart, breathing and digestive system. If anything untoward happens – the fire alarm goes off or a wasp stings you – it will trigger a host of appropriate responses. The entire control mechanism provided by your nervous system (brain, spinal cord and nerves) is the subject matter of neuroscience.

The story of neuroscience began in prehistoric times, though it only really became 'neuroscience' in the last hundred years or so. Its remit is from the study of the cellular- and molecular-level action of individual neurons (nerve cells) to understanding how entire neural systems work to produce movement, sensation and cognition.

Right at the heart of neuroscience lies an intractable problem: somehow, the physical and chemical processes of the brain and nerves create the myriad intangible effects of consciousness, thought, imagination, memory, intention, emotion, personality. But how? How does human experience emerge from a cluster of biochemical processes? How is the mental intention to do something translated into physical movement, or the impact of a stimulus such as a sight or sound translated into enjoyment or anguish which seems not to be located in any part of the body?

Neuroscience is a new term and a new discipline. The first few millennia of our

MEET YOUR BRAIN

The brain lies within the skull, protected by membranes called the meninges. It has three main parts. The largest is the cerebrum, which is divided into two symmetrical halves and is deeply folded. Its outer layer, called the cerebral cortex, has discrete functions such as dealing with input from the senses, control of physical actions and higher mental functions such as language and abstract thought. The smaller cerebellum, at the back of the head, is important in motor control, balance and coordination. The third part is the brain stem, which is responsible for moving information between brain and body. The material near the outside of the brain is grey and that inside is whitish. The grey matter is made up of the cell bodies of brain cells (neurons), and the white matter consists of bunches of axons (nerve fibres) which connect them.

narrative must necessarily be drawn from other disciplines, including philosophy, physiology, physics, chemistry and other sciences. From these we can uncover emerging understanding of how our sensory systems work, how we control our bodies, and how learning and memory operate. But our understanding is far from complete – the story of neuroscience is a narrative that is still unfolding.

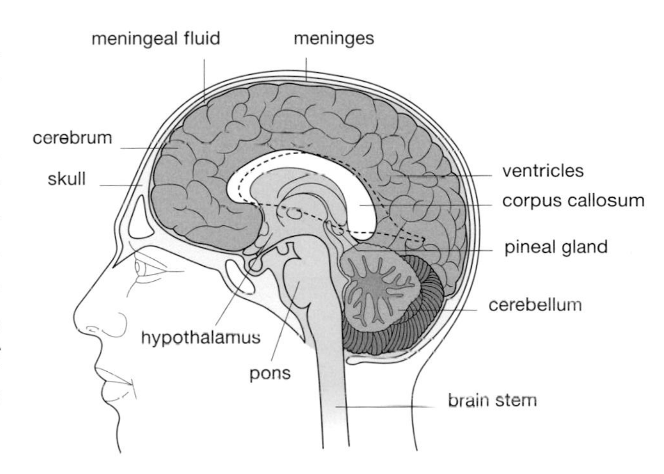

Right: important areas and structures of the human brain.

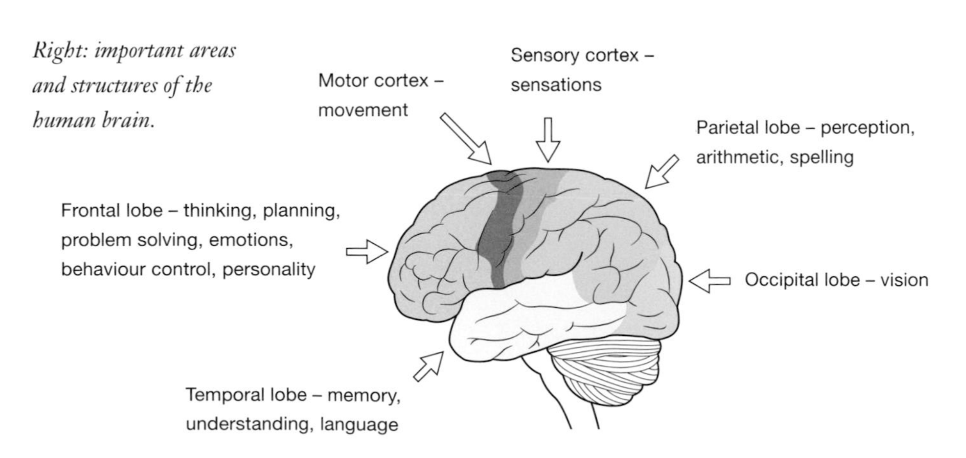

CYRUS

CHAPTER 1

Who's in CONTROL?

'Tell me where is fancy bred,
Or in the heart or in the head?'

William Shakespeare,
The Merchant of Venice,
Act III, Scene 2

It's clear that something in the body is responsible for coordinating all it does. Some part of us obviously controls sense perceptions, motion, automatic functions such as breathing, and provides the emotional and intellectual activity we assign to our minds. But there's little to link these functions with the brain or even to suggest they are all carried out by the same organ. For this reason, it was not immediately obvious to our ancestors that the brain performs these functions.

An athletic manoeuvre such as this demands a lot of the brain as well as the body.

Heart v. brain

Many early cultures associated emotions and thought with internal organs. But there is no physical evidence in the body that helps us to locate emotions, personality or consciousness, so they have been linked with different body parts by different cultures. In Mesopotamia, 4,000 years ago, the heart was thought to house the intellect, while the liver was considered the centre of thought and feeling, the womb of compassion (obviously men were not compassionate) and the stomach of cunning. In Babylonia and India, too, the heart was king.

First brains

The Ancient Egyptians were at one point aware of the brain's importance in controlling the body. The earliest known medical text is the Edwin Smith papyrus, produced around 1700BC but probably based on material 1,000 years older. It provides descriptions of 48 cases of injuries, aiming to guide the surgeon in determining whether to attempt to treat a patient. The surgeon realizes that if the neck is broken the patient can become paraplegic or quadriplegic as the connection between the brain and limbs is

The Edwin Smith papyrus preserves Egyptian medical lore which probably dates from around 2700BC.

lost and cannot be restored. The papyrus provides the first ever description of the human brain. It is said to be like 'those corrugations which form in molten copper' and that the surgeon might feel something 'throbbing' and 'fluttering' beneath his fingers like 'the weak place of an infant's crown before it becomes whole'.

Even so, the Egyptians were so certain the brain was not a vital organ that they hooked it out through the nose and discarded it when mummifying a corpse, but preserved other organs in canopic jars. Like several early civilizations, the Egyptians considered the heart to be the centre of the intellect and home of the mind.

Perhaps it is hardly surprising that the complex role of the brain was obscure. A bit of post mortem examination reveals the approximate function of most of the major organs. The heart is connected to the blood vessels, the kidneys to the urinary bladder, the gut connects the mouth and the anus by a circuitous route – but it is not at all clear what the brain is *for*.

Champions of the brain

The brain was first promoted as the seat of the intellect by Ancient Greek philosopher Alcmaeon of Croton in the 5th century BC. He is the first person known to have carried out dissections with the intention of finding out how the body works. He dissected the optic nerve, and wrote of the brain as the centre of processing sensations and composing thought. Around the same time, medical writer Hippocrates also assigned considerable power to the brain: 'I am of the opinion that the brain exercises the greatest power in the man. . . . The eyes, the ears, the tongue and the feet, administer such things as the brain cogitates. . . . It is the brain which is the messenger to the understanding.' However, this was by no means the only or predominant view in Ancient Greece.

> *'The seat of sensation is in the brain. This contains the governing faculty. All the senses are connected in some way with the brain. . . . This power of the brain to synthesize sensations makes it also the seat of thought: the storing up of perceptions gives memory and belief, and when these are stabilized you get knowledge.'*
>
> Alcmaeon of Croton, 5th century BC

The lusty liver

The pre-Socratic Greek philosopher Democritus (460–371BC) divided the functions we now assign to the brain between three organs. He attributed consciousness and thought to the brain, emotions to the heart and lusts and appetites to the liver. Plato (428–347BC) later developed this idea into the three-part soul (see page 18), locating reason or intellect in the brain, which he declared to be 'the divinest part of us, and lords it over the rest'.

Hippocrates' treatise on epilepsy, *On the Sacred Disease*, written about 425BC, cites the brain as the source of pleasure, grief and all other feelings. He says the heart makes sense perceptions and judgement possible and is also the site of madness, delirium and terror and of the causes of insomnia and poor memory.

The Greek philosopher Democritus located consciousness in the brain.

MATTER MATTERS

Democritus taught that all matter is made up of tiny 'uncuttable' portions called atoms and that the different qualities of matter come about through the combination and configuration of the different types of atoms within it. The most refined matter was, in his model, made up of the smallest spherical atoms. The psyche (soul or mind) was made up of these refined atoms and concentrated in the brain. Larger and slower atoms predominated in the heart, which he considered the centre of the emotions, and still cruder atoms in the liver, the home of the appetites.

Brain and nerves

The first anatomists to carry out a detailed study of the human brain and to dissect it were Herophilus (c.335–280BC) and Erasistratus (304–250BC) in Alexandria, Egypt. They are said also to have carried out vivisection on human prisoners, a practice defended by the Roman writer Celsus in the 1st century AD: 'Nor is it cruel, as most people maintain, that remedies for innocent people of all times should be sought in the sacrifice of people guilty of crimes, and only a few such people at that.'

Herophilus is credited with the discovery of the nerves, being the first to distinguish between nerves, blood vessels and tendons (which all look rather similar). It's possible that he and Erasistratus were aware of the distinction between motor and sensory nerves (see page 86); certainly, Herophilus was aware that damage to some nerves could cause paralysis. They, too, regarded the brain as responsible for thought and sensation, distinguished between the cerebellum and cerebrum, and named both the meninges (membranes surrounding the brain) and the ventricles (spaces filled with cerebrospinal fluid). Herophilus recognized the brain as the centre of the intellect, and placed the command centre in the fourth ventricle. He likened the cavity in the posterior floor of the fourth ventricle to the reed pens used in Alexandria. The cavity is still called the *calamus scriptorius* ('reed pen') or *calamus Herophili.*

Resurgence of the heart

It might seem that the scene was now set for a steady understanding of the function of the brain to emerge, but unfortunately an influential thinker took a different view. The philosopher Aristotle (384–322BC) was convinced that the heart was the 'command centre' of the body, responsible for sensation, movement and psychological activity, whereas the brain served only as a

Herophilus and Erasistratus are the first people known to have worked on the nerves, shown here in a drawing from 1532.

cooling chamber of some kind. He argued against the hegemony of the brain on several counts, most of them inaccurate:

- The heart is connected to all of the rest of the body via blood vessels, whereas the brain has no comparable connections (it has – but nerves are difficult to see in a dissection using primitive instruments).
- Not all animals have a brain (almost all do, but some invertebrates don't).
- The heart develops before the brain in an embryo.
- The heart provides blood which is needed for sensation, while the brain has no blood supply (neither of these is true).
- The heart is warm but the brain is cold (they are about the same temperature).
- The heart is essential for life, but the brain is not (true in some primitive animals).
- The heart is sensitive to touch but the brain is not, and the heart is affected by emotions.

As we shall see (page 31), Aristotle rejected the notion of a metaphysical entity such as a spirit, soul or mind which inhabits a body and is separable from it. He believed the '*pneuma*' or life-breath which animates the body is entirely material and expires on the death of the organism. This meant he had to locate all psychic function in the physical body – and he chose the heart.

The Stoic philosophers of the 3rd century BC favoured Aristotle's position,

> *'The brain is not responsible for any of the sensations at all.'*
>
> Aristotle, 4th century BC

The Ancient Egyptians' reverence for the heart as the seat of wisdom and the soul lay behind the motif of the 'weighing of the heart'. After death, two gods, Thoth and Anubis, weigh the heart of the deceased to determine the person's worthiness (c.984BC).

arguing that speech is associated with thought and breath, and as speech rises through the windpipe it must originate in the chest and therefore the thought that leads to it must also come from the chest. (This might seem odd reasoning, but later the observation that the eyes, ears, nose and mouth are all in the head, close to the brain, was considered good support for the notion that sensory input is dealt with by the brain.)

Galen's gladiators

In the 2nd century AD, the Roman physician Galen was convinced that the brain was the most important organ in terms of controlling the body. He was in a much better position than Aristotle to make an informed decision. Galen was a surgeon and often treated gladiators who suffered a wide variety of devastating injuries. He soon discovered that severing the spinal column deprived parts of the body below the injury of both sensation and movement. He noted that the extent of damage to respiration, speech and other functions depended on the location and extent of injuries to the nerves and muscles. He learned to distinguish between motor and sensory nerves in terms of their appearance as well as their function, and traced their connection to the spinal cord and brain.

The pig decides it

Galen's insistence that the brain was the control centre of the body was a little awkward, as his most important patient was the Stoic philosopher and Roman Emperor Marcus Aurelius, and the primacy of the heart was widely accepted by the Stoics.

Undeterred, Galen contrived a public demonstration to show conclusively that the brain controls the muscles via the nerves. The demonstration involved an unfortunate pig. (There were many unfortunate animals, usually pigs or macaques, involved in Galen's experiments and demonstrations.) This particular demonstration arose after Galen accidentally cut the laryngeal nerves (nerves to the larynx or voice box) during an exploration of breathing. The pig involved was tied down and squealing (understandably) while he operated on it. When Galen cut the nerve, the pig continued to struggle, but stopped squealing. Investigation revealed that Galen had severed the nerve connecting the larynx to the brain. As the experiment involved an animal rather than a human patient, it was repeatable. Galen organized public demonstrations in which he cut the laryngeal nerve of a secured pig, thus silencing both the pig and his opponents.

The squealing pig became one of the most famous physiological demonstrations of all time, and was the first experimental evidence that the brain controls behaviour. When one of the rhetoricians who had originally challenged Galen said that he had proved only that the brain controls the squealing of the pig and not the rationality of the human speaker, Galen responded that he had seen the laryngeal nerve accidentally severed during an operation on a human patient and it had the same effect of destroying the power of speech. This 'accident' seems quite a lucky coincidence for Galen, if his account is true.

But the question was not resolved quite so easily. It is still a widespread

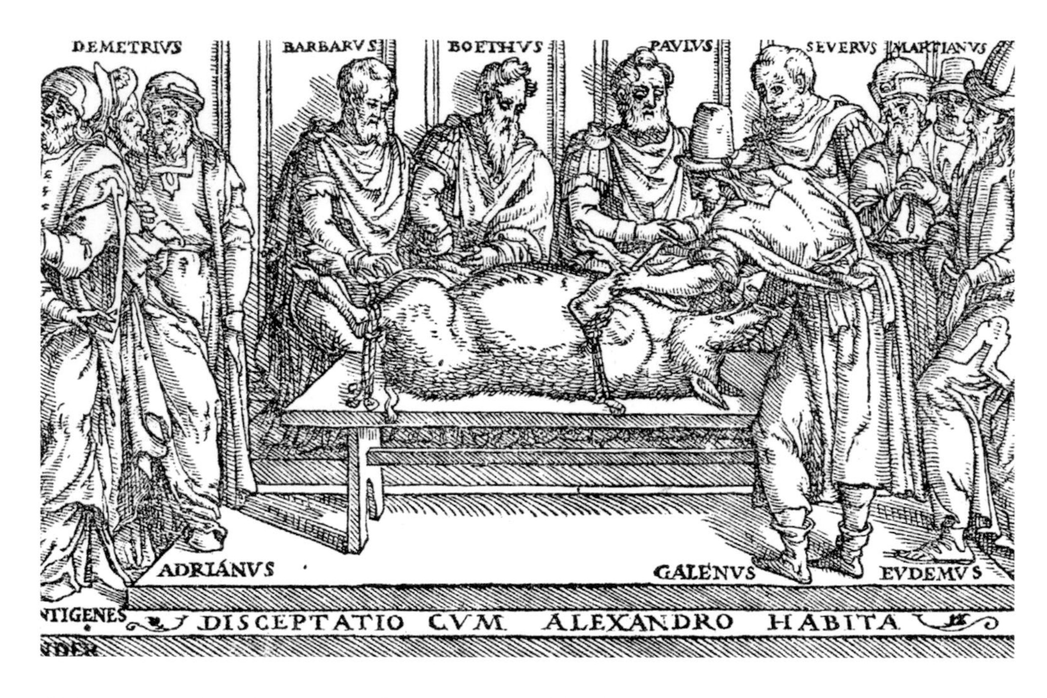

Galen preparing for a demonstration that never went well for the pig.

conceit that the heart is the origin of strong emotions. Although the opinions of Plato and Galen dominated thinking in the Arab world into the Middle Ages, a parallel trend promoting control by the heart also continued. Some authorities even considered the responsibility divided between heart and brain. The Arab physician Ibn Sina (980–1037), also known as Avicenna, held the brain responsible for cognition, sensation and movement, but thought it was controlled by the heart. It was as though the heart were delegating major elements of the body's management to the brain. Most animals do not fare well if either the heart or the brain is removed, so it was difficult to prove by experimentation whether one or both were necessary for movement and cognition.

Ibn Sina's Rules *about medicines of the heart.*

GALEN, (AD129–c.200)

Galen was born into an upper-class family in Turkey, then part of the Roman Empire. He began to study medicine at the age of 16 after his father dreamt that he had to change the course of his son's study from mathematics and philosophy. Galen successfully trained as a doctor, but worked more in research than clinical practice until he was 28. Having already produced several books, he was then appointed physician to the gladiators in his home town of Pergamon. This would have given him considerable experience of dealing with traumatic injuries.

In 161, war closed the gladiators' school and Galen moved to Rome. He became immensely successful and highly regarded in society and was appointed private physician to three emperors in turn. He had many arguments with other physicians and philosophers, and wrote extensively on physiology, medicine and anatomy.

Galen's work on physiology and anatomy was grounded in practice and detailed observation. He was the most accomplished medical professional and thinker of the classical world and his work dominated medical practice and physiology until the 16th century. However, Galen's dissections were carried out on animal subjects and many of the descriptions and conclusions he drew did not apply to human anatomy. Such was the regard in which his work was held, though, that even these gross errors went unchallenged until the time of the Renaissance anatomists.

The three great ancient teachers of medicine: Galen (Roman), Avicenna (Persian) and Hippocrates (Greek), shown in an early 15th-century medical text.

Brain, nerves and 'soul'

The demonstration with the pig was physiological: cutting the nerve prevented the vocal cords from working. In Galen's model of how the nerves worked, it would have had this effect by preventing the flow of *pneuma* (see page 14) through them. But Galen believed that the nerves and brain performed more than a simple mechanical function. He distinguished a motor soul and a sensory soul. The sensory soul had five attributes, equivalent to the five perceptual faculties, but the motor soul had just one: movement. Galen also believed the rational soul to have three functions – reason, imagination and memory. In this, then, he identified three principal functions of the brain: perception and sensation; controlling movement; and psychological activity.

The matter of the soul

Aristotle believed that the *pneuma* entered as breath through the bronchioles in the lungs, travelled by the pulmonary vein to the heart and was there converted to 'vital *pneuma*'. This, in turn, was carried by the blood vessels to the muscles, making them contract. Like Aristotle, Galen believed in a vital *pneuma* made by the body from components drawn from food consumed and air breathed in. In Galen's model, the most basic *pneuma* is made in the liver, where the products of digestion mix with blood to infuse it with natural spirits. The supplemented blood goes to the heart where impurities are removed and extra spirits drawn from the lungs are mixed with it, forming the next stage of *pneuma* – vital spirits. From the heart, the enriched blood flows to a network of blood vessels around the base of the brain called the *rete mirabile* where it is further enriched, becoming the highest form of *pneuma* – psychic spirits. In fact, the *rete mirabile* does not exist in humans; it was something Galen had seen in dissections of the ox brain and he assumed

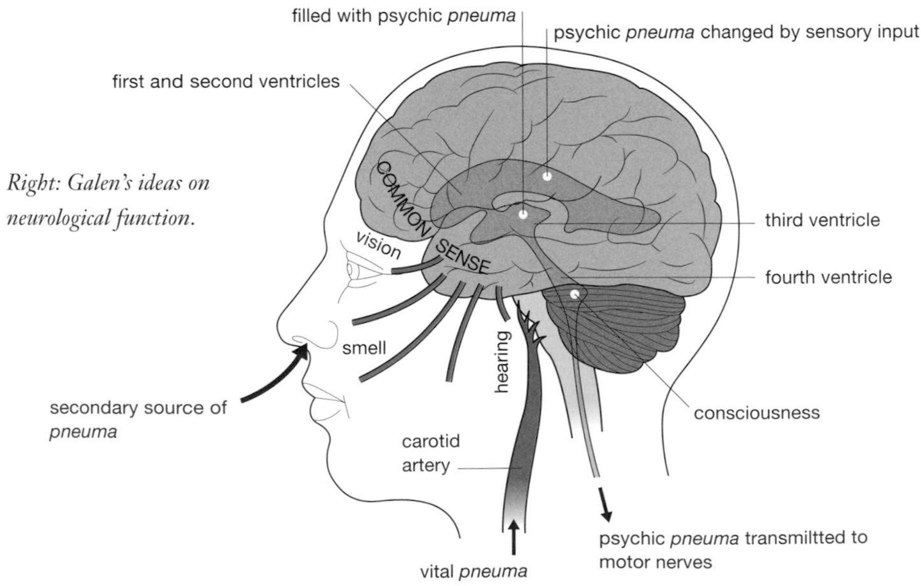

Right: Galen's ideas on neurological function.

LOCATING REASON

Before Galen located the ventricles in the brain as the area responsible for reason, various philosophers had held it to be in different places:

- Plato made the entire head responsible for reason.
- Strato sited reason in the space between the eyebrows.
- Erastistratus located reason in the meninges, the membranes around the brain.
- Herophilus held the ventricles of the brain to be responsible for reason.
- Parmenides favoured the thorax.
- Aristotle put reason in the chest.
- Empedocles thought the blood within and around the heart provided reason.

it was also present around the human brain. Still, since psychic *pneuma* doesn't exist either, it's of no great consequence. According to Galen, the psychic *pneuma* then flows into the ventricles of the brain from where it can either be sent out to areas of the body to have an effect on the muscles, or used in the brain to effect mental activity.

A good starting point

By the time of Galen's death, there was good reason to consider the brain to be the control centre of the body, that it communicates with the rest of the body by means of sensory and motor nerves, which are distinct, and is also the source of much, if not all, mental activity.

The real problem was that there was no evidence to show how it all worked. Without a grounding in physiology, the question of what the brain and nerves do, and how they do it, was approached from philosophical beliefs or assumptions. The foundation of centuries of philosophizing was then based on the work of Galen, rooted not in detailed examination of human anatomy but in animal dissections and conjecture. In later centuries, emphasis shifted to trying to ground theory about what the brain and nerves do in observations of their structure.

Looking at the brain

The first people deliberately to lay the brain open to scrutiny were based in Kos in Greece around 300BC. Praxagoras, who was the first to distinguish between arteries and veins, described the 'long flexuosities and winding and folding' convolutions of the cerebral cortex. But this does not give any indication of what the convolutions or 'flexuosities' might do.

Erasistratus compared the brains of several other animals with those of humans, finding the human brain to have the most convolutions. He concluded that superior intelligence accompanies a more convoluted brain – an idea later ridiculed by Galen who pointed out that a donkey's brain has more convolutions than a human brain. The huge and complex surface of the brain, the cortex, was thereafter ignored. That wasn't a good start.

Starting with the gaps

Galen was the first to localize function within the brain. He argued that the sensory nerves go to the front of the brain, and that both the nerves and that part of the brain share the same nature, being soft

> *'Even asses have a complex and convoluted brain while, because of their stupid temperament, these animals should have a very simple and uniform brain.*
> *. . . In my opinion, the degree of intelligence does not depend on the quantity of the psychic pneuma, but on its quality.'*
>
> Galen

and impressionable, and therefore suited to dealing with perception. The 'harder' motor nerves he believed had their root in the back part of the brain and extended from there throughout the body, so the anterior portion of the brain was, he concluded, concerned with movement.

Tracing the bundles of nerves that enter the brain from the eyes, ears, nose and mouth and from the spinal cord gave a clear indication of how the brain could be involved in receiving sensory input and controlling movement of the body. But it did not help to explain that other, more nebulous function of the brain: the psychological activity which makes us human and individual. There are no physical structures that clearly relate to mental activity. Galen attributed the rational soul to the whole of the brain, but with the higher faculty of reason located in the ventricles, which he considered to be filled with vital *pneuma*.

The three-cell model of the brain

In the early years of Christianity, the Church fathers struggled to accommodate the idea of the incorporeal soul within the physical body (see pages 33–5). They linked the soul with cognitive function but did not want to locate it in the solid structure of the brain, so the fluid-filled ventricles seemed the most appropriate location for mediation between the solid body and the insubstantial soul.

The Greek Christian philosopher and bishop Nemesius developed a doctrine of ventricular localization around AD390. In this, he proposed that the brain comprises three cells which correspond to the ventricles (with the lateral ventricles combined into a single cell), each responsible for a different type of cognitive or perceptive ability.

Following Galen's belief that sensory perceptions are received at the front of the brain, Nemesius located sense perception in the first cell. Here, he believed, perceptions are processed through 'common sense' (*sensus communis*) to yield understanding of the object perceived. This is achieved by bringing together the input from all the senses, with the soul acting on the data to produce a unified impression. For example, combining appearance, feel and smell produces recognition and understanding of an orange. Since vision involves producing images, other processes involving vision, including fantasy and imagination, must also be located in this first cell, which

> *'Since the emptying of the* pneuma *from the hollows of the brain, when it is wounded, at once makes men both motionless and without feeling, it must surely be that this* pneuma *is either the very substance of the soul or its primary organ.'*
>
> Galen, *On the Use of Breathing*

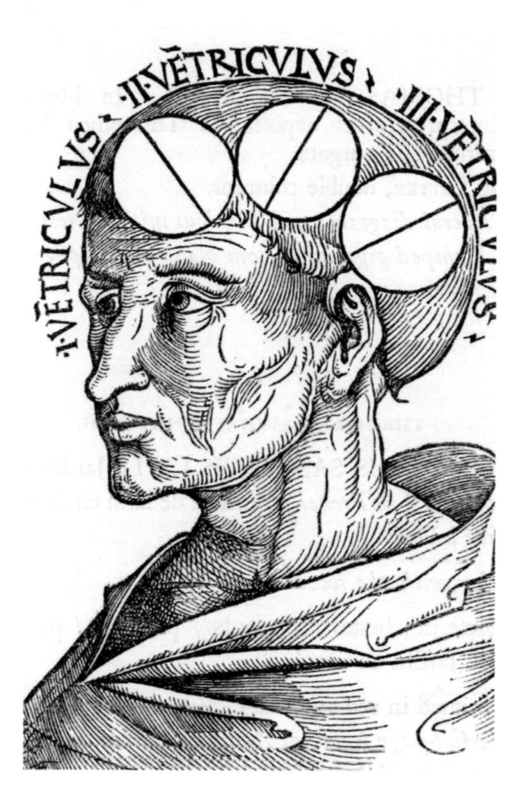

The three-cell model of the brain, from an edition of Philosophia Naturalis *by Albertus Magnus, published in 1506.*

became known as the *cellula phantastica*. The second cell was considered the site of reasoning, thought and judgement. This was the *cellula logistica*. The final cell, the *cellula memorialis*, was where knowledge was stored as memory. There is, then, a clear flow of information in psychic *pneuma* through the *foramen* (the holes leading from one ventricle to the next), giving a progression from perception through cognitive processing to stored knowledge or memory, the information moving from the front of the brain towards the back.

Nemesius claimed supporting evidence from observing the impact of damage to the brain. Lesions in the front ventricles, he claimed, impair sensory perception but not intellect. Damage to the central part of the brain causes mental derangement but doesn't affect sensory perception, while damage to the cerebellum causes loss of memory, but does not impair perception or thought processes. This doctrine of ventricular localization, as it became known, was accepted without challenge for around 1,000 years.

The 12th-century writer Master Nicolaus (possibly of Salerno) went a step further, itemizing the characteristics of the three cells in terms of humoral theory (see box on page 23). He thought they must have varying amounts of 'marrow' and 'spirits', which reflected their conditions in terms of the humours, being hot or cold and wet or dry.

According to Nicolaus, the *cellula phantastica* is hot and dry and abundant in spirit. Heat and dryness attract the animal spirit, and the presence of the spirit helps the flow of spirit carrying information. There is little marrow, as this would impede the flow of spirit and apprehension of the nature of things. The *cellula logistica* is hot and moist because this provides for discrimination, allowing the brain to process those ideas passed on from the *cellula phantastica*, distinguishing which are true rather than false, honest rather than dishonest, and so on. Again, there is abundant spirit as that is needed for the cell's activity, but also much marrow, which replenishes the spirits when they become depleted. The *cellula memorialis* is cold and dry, because these

Constructing the object 'orange' from sensory data is a complex mental task.

> *'[The brain] is divided into three cells, the* cellula phantastica *in the anterior part of the head, the* cellula logistica *in the middle, the* cellula memorialis *in the posterior part. In the* cellula phantastica, *imagination is said to have its seat, reason in the* cellula logistica, *memory in the* cellula memorialis. . . . *First we gather ideas into the* cellula phantastica, *in the second cell we think them over, in the third we lay down our thoughts; that is, we commit to memory.'*
>
> The Anatomy of Master Nicolaus, c.1150–1200

IN THE BRAIN BUT NOT OF IT

Nemesius was a Christian, and his concept of the soul was in keeping with his religion rather than with the Aristotelian model. He did not consider the soul to be an integral part of the body, but a distinct and separate substance that mixed with the body or inhabited it for the duration of life. The acts of emotion, thinking, perception and so on were, for him, acts of the soul and not of the brain. It is an important distinction, and one that continues to have an impact on neuroscience.

properties help retention. It has abundant marrow so 'that it may be easily stamped with the impressions of diverse ideas, but not much spirits, which might flow about and remove the impressions of ideas'.

Cells, spirits and senses

It is quite clear that these notions of the cells, their nature and their function were not deduced from observing either the dead brain or the active brain of an animal. The entire construct is built on the philosophical notions of how it was thought the brain *should* work, and how it could be fitted into the existing models of a body controlled by humours and a brain which housed mental capacities in the ventricles.

Several Arab and early Christian writers extended the number of internal senses from three to five and then seven. There remained only three cells in which to locate them, so they had to share the space. Some authors gave precise locations for each, for example, splitting one or more of the cells into top, middle and bottom locations.

Some writers extended the number of cells to four; all considered the different

HUMOURS OF BODY AND SOUL

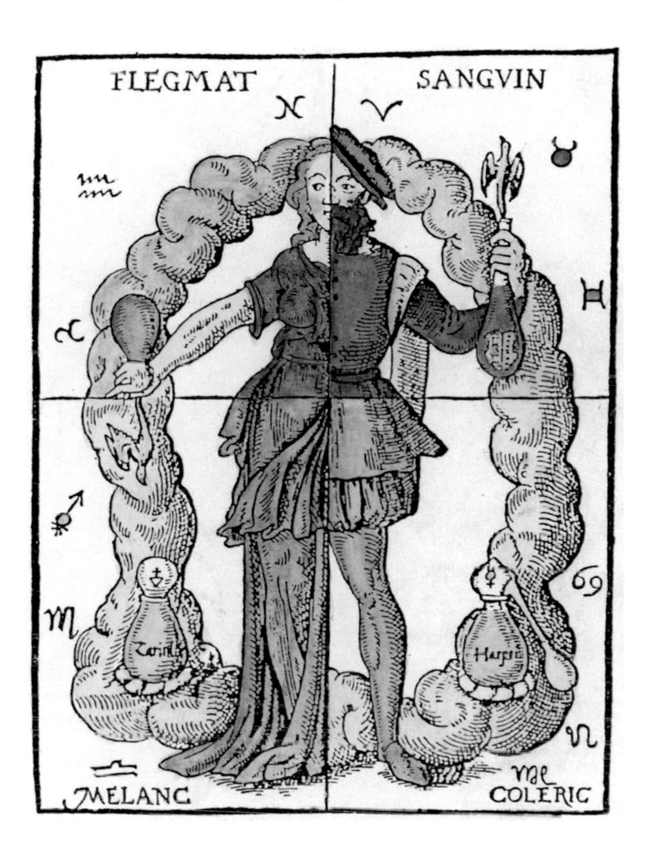

An idea originating in Hippocratic medicine in the 5th century BC had the body governed by different 'humours' corresponding to four fluids. The humours were related to the four types of matter, or 'roots', set out by Empedocles in the 5th century BC. The roots of matter were earth, water, air and fire and they had properties of heat/cold and wetness/dryness.

Corresponding to these in the human body, Hippocrates described the four humours identified with four body fluids: melancholy (black bile), choleric (yellow bile), sanguine (blood) and phlegmatic (phlegm). Human health, he taught, relied on keeping the humours in the correct balance. According to the theory, each individual has their own natural balance of humours and this dictates their temperament as well as health.

internal senses to be executed by distinct types of spirit. The treatment of the spirits/faculties and their locations was more precise in the philosophical texts than the medical texts – tellingly, as the philosophical texts neither needed nor had any anatomical foundation. The medical texts, on the other hand, assumed that damage to one cell of the brain would (or could) affect all the faculties resident there. When injuries or disease did not trigger the expected symptoms, the entrenched dogma usually triumphed over observation. So when the French surgeon Guy de Chauliac (1300–68) investigated a patient with severe damage to the posterior ventricle, but no loss of memory, he assumed that the damage was insufficiently severe to cause memory loss but failed to question the validity of the localization of memory.

Leonardo and the brain

The three-cell model of the brain went unchallenged until the 16th century when rigorous human dissection began in Renaissance Europe. Or almost: at least one person had found fault with it a little earlier, but his ideas never escaped his own private notebooks to gain wider currency.

The great scientist and polymath Leonardo da Vinci (1452–1519) had a particular interest in how the brain processes sensory information and passes it to the soul. He dissected and drew the brain, eventually producing more detailed and strictly anatomical depictions than anyone previously. His first drawings relied heavily on Arab interpretations of Galen and on medieval tradition. In keeping with this, his depiction of the visual system shows a connection from the eye to the first 'cell'; he also connected the ears to the first cell.

Several years after his first drawings of the brain, Leonardo began to make wax casts to discover the shape of the ventricles. This worked well. He described

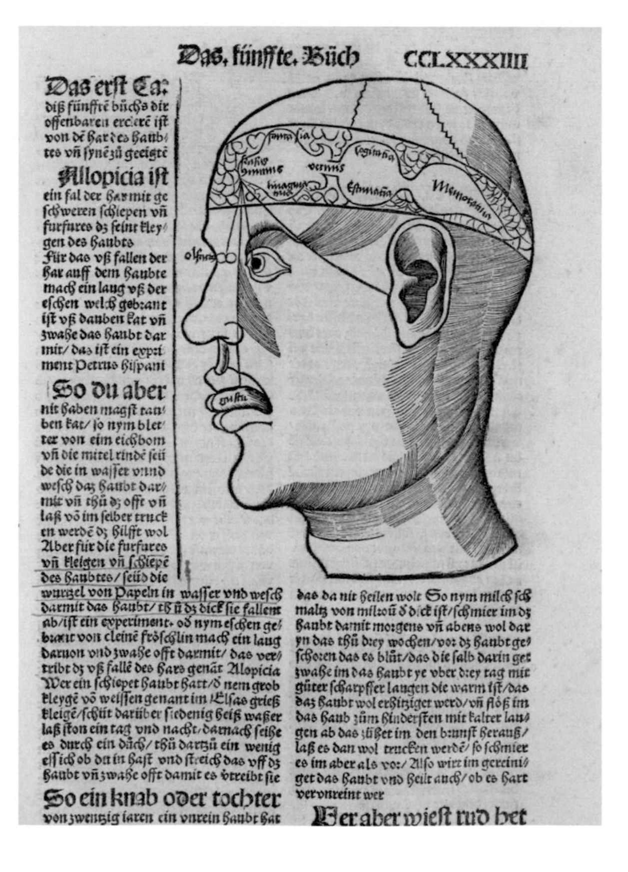

Das. fünffte. Büch CCLXXXIIII

Das erst Ca:
diß fünfftē büchs dir offenbaren ercleré ist von dē har des haub-tes vñ synē zü geeigtē

Allopicia ist
ein fal der har mit ge schweren schiepen vñ furfures dz seint kley-gen des haubts
Für das vß fallen der har auff dem haubte mach ein laug vß der eschen welch gebrant ist vß dauben kat vñ zwahe das haubt dar mit/ das ist ein expri ment Petrus hispani

So du aber
nit haben magst tau-ben kat/ so nym bler-ter von eim eichbom vñ die mitel rindē seü de die in wasser vnnd wesch daz haubt dar-mit vñ thü dz offt vñ laß võ im selber truck en werdē dz hilfft wol
Aber für die furfures vñ kleigen vñ schiepē des haubtes/ seüd die wurzel von Papeln in wasser vnd wesch darmit das haubt/ thü dz dick sie fallent ab/ ist ein experiment. od nym eschen ge-brant von cleinē fröschlin mach ein laug darvon vnd zwahe offt darmit/ das ver-tribt dz vß fallē des hars genāt Alopicia
Wer ein schiepet haubt hatt/ d nem grob kleygē võ weissen genant im Elsas grieß kleigē/ schüt darüber siedenig heiß waßer laß ston ein tag vnd nacht/ darnach seihe es durch ein düch/ thü dartzü ein wenig essich ob du in hast vnd streich das vff dz haubt vñ zwahe offt damit es vtreibt sie

So ein knab oder tochter
von zwentzig iaren ein vnrein haubt hat das da nit heilen wolt So nym milch sch maltz von milrou d dick ist/ schmier im dz haubt damit morgens vñ abens wol dar yn das thü drey wochen/ vor dz haubt ge-schoren das es blüt/ das die salb darin get zwahe im das haubt ye vber drey tag mit güter scharpffer laugen die warm ist/ das daz haubt wol erhitziget werd/ vñ flöß im das haub züm hindersten mit kalter lau-gen ab das züßet im den brunst herauß/ laß es dan wol trucken werdē/ so schmier es im aber als vor/ Also wirt im gereini-get das haubt vnd heilt auch/ ob es hart verunreint wer

Wer aber wiest rud het

Localization of functions in the three cells of the brain, Gregor Reisch, Margarita Philosophica, *1503.*

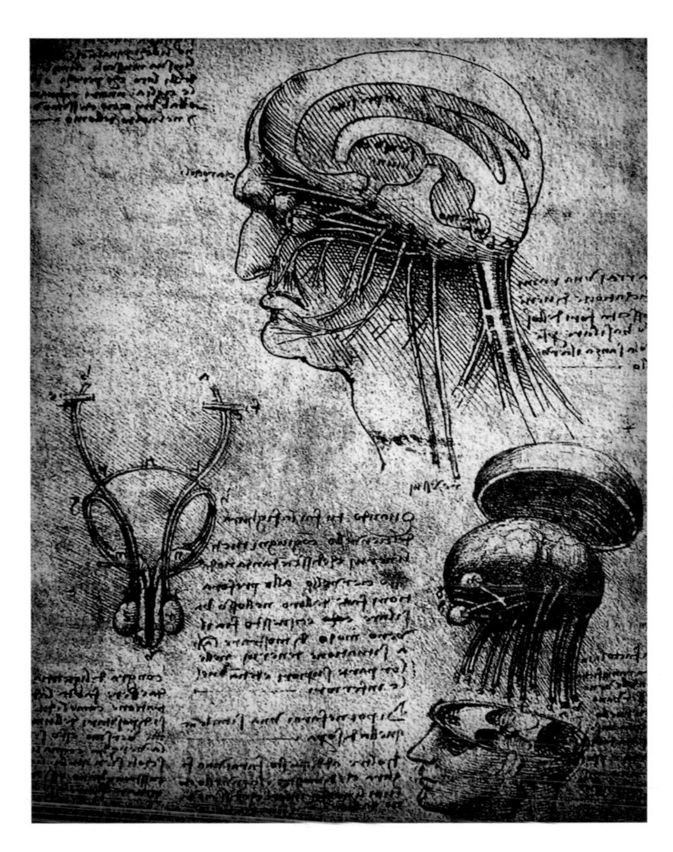

In this diagram, Leonardo da Vinci presented the first ever transparent anatomical cross-section (top) and the first exploded diagram (bottom right).

of the hand. He finally broke with the tradition of locating the *sensus communis* in the first cell, moving it to the second cell (third ventricle). It seems he was prompted to make the change after finding that the trigeminal nerve (responsible for sensation in the face and for chewing activity) and the auditory nerves (responsible for hearing) terminate here. He thought all tactile input went to the third cell (fourth ventricle), which rather confused his system. Indeed, he gave up describing, and left the nerves to tell their own story in his later diagrams. Sadly, Leonardo did not publish his anatomical work, so it had no impact on his contemporaries.

a technique for making holes in the ventricles (of a corpse), inserting narrow tubes into the holes to allow air/liquid to escape, and then filling the ventricle with molten wax through a syringe. After the wax hardened, he removed the brain tissues to reveal his cast of the ventricle. The same technique was not used again to model spaces in the body until the 18th century, when Dutch anatomist Frederik Ruysch rediscovered it.

Leonardo carried out many dissections tracing the paths of nerves, particularly from the eyes, nose and mouth to the brain, and including the vagus nerve which connects with the abdomen, and the nerves

> *'A foolish extravagant spirit, full of forms, figures, shapes, objects, ideas, apprehensions, motions, revolutions: these are begot in the ventricle of memory.'*
>
> William Shakespeare, *Love's Labours Lost*, Act IV, Scene 2

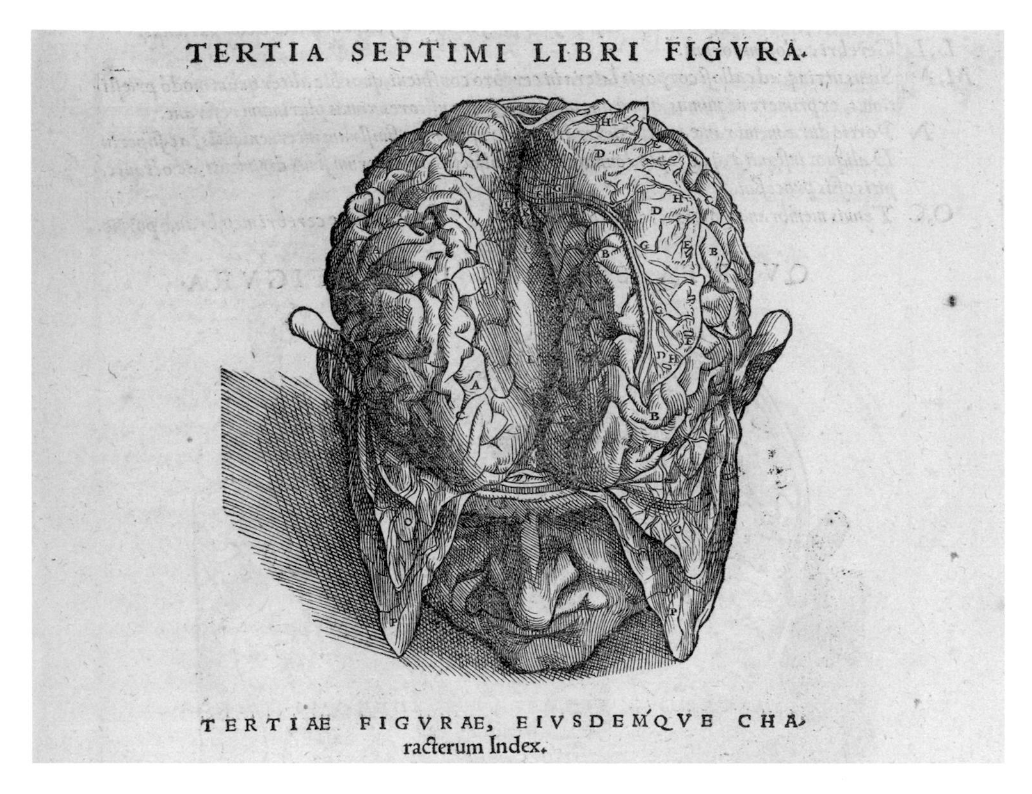

In De humani corporis, *Vesalius published intricate diagrams of the dissected brain. Here the skin is peeled back to reveal the cerebral cortex and the brain split to divide the hemispheres.*

Remaking the brain

The three-cell model was finally challenged (ridiculed, even) by the great Flemish anatomist Andreas Vesalius (1514–64). He pointed out that many animals have ventricles, but we do not endow them with a soul, so the human soul cannot reside in the ventricles.

> *'All the nerves manifestly arise from the spinal cord . . . and the spinal cord consists of the same substance as the brain from which it is derived.'*
>
> Leonardo da Vinci, notebooks

Vesalius worked from human dissections which he carried out himself. He rejected animal vivisection as he considered it was wrong to rob animals of their cognitive abilities, even if those were inferior to the abilities of humans. His decision to work solely with human subjects meant he was well-placed to signal some of the errors that had persisted since the time of Galen and were attributable to Galen's practice of examining the brains of oxen and macaques.

Vesalius is the first person to have published intricate diagrams of the dissected

brain. They appeared in his groundbreaking work, *De humani corporis fabrica* (*On the workings of the human body*), in 1543. He did not draw the illustrations himself but commissioned an artist (possibly Jan van Calcar) to do them, working under his direction during dissection.

Vesalius' work was vital in shifting the perspective on studying the brain from endorsing philosophical views already held to comprehending anatomical discoveries. This was in the spirit of the European Renaissance, a time when scientists finally began to challenge the authority of classical writers and, at least in some regards, question the dogma of the Christian church.

Getting real

During the 16th and 17th centuries, dissection revealed far more about the structure of the brain, but still there was little information forthcoming about its operation or how form relates to function. Then, in the 1660s, two anatomists working independently set out to challenge the authority of Galen. The English physician Thomas Willis published his *Anatomy of the Brain* in 1664 and the Danish anatomist Nicolaus Steno published *Lecture on the Anatomy of the Brain* in 1669.

Willis claimed that the ventricles were formed 'accidentally from the complication of the brain', and were not part of God's plan to produce a cosy location for the soul. He did important work in revealing the structure of the brain. In particular, he injected ink into the blood vessels, which allowed him to trace their path around and through the brain. Steno was bold enough to state that the *sensus communis* does not exist; he had made a careful study of the ventricles and found no sign of it. He rejected Galen's theory of animal spirits.

It is here, with the consensus in favour of the brain controlling physical and mental functions, and with it communicating with the body by some unknown means (but not animal spirits) that the beginnings of modern neuroscience can be found.

HEART-OR-BRAIN REVISITED

Eventually, the question of whether the heart or brain is in control, at least of motor activity, was settled by experiment. In 1664, Dutch microscopist Jan Swammerdam (1637–80) carried out a rather gruesome demonstration to prove to the Danish botanist Olaf Borch that the heart is neither the source of movement nor necessary for nerve transmission to the muscles.

Swammerdam cut the heart from a living frog and set the animal back in the water, where it continued to swim for a while – not happily, one imagines, but successfully. When he repeated the experiment, but removed the brain of the frog, it could no longer swim. If any doubt remained about the action controlling the muscles originating in the brain, it was quashed by Swammerdam's unfortunate frog.

Die Seele deß Menschen

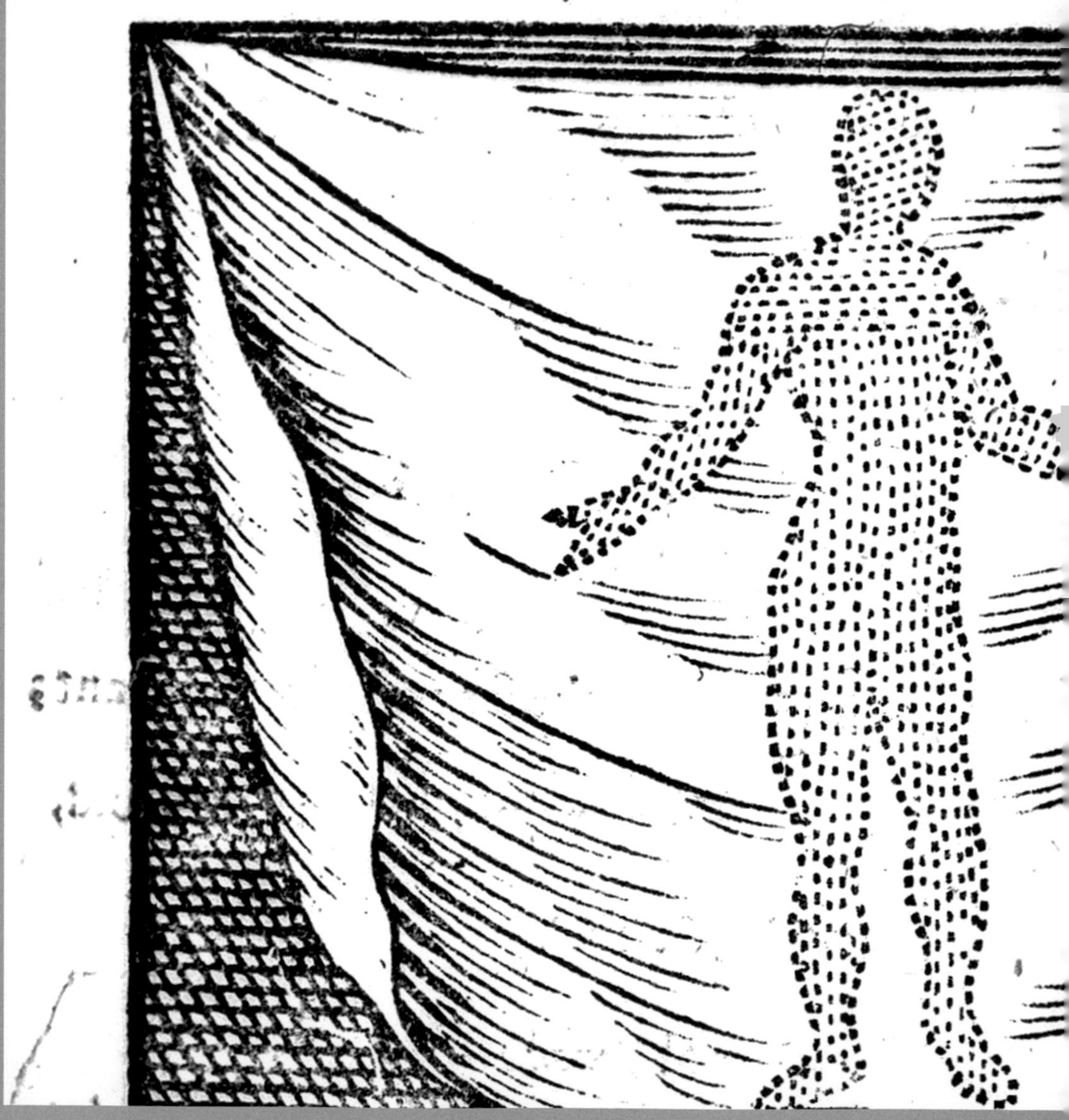

CHAPTER 2

The ghost in the MACHINE

'Man is compounded of two very different ingredients, spirit and matter, but how two such unallied and disproportioned substances should act upon each other, no man's learning could yet tell him.'

Samuel Johnson, 1755

We are all aware of some inner aspect of ourselves which we associate with thought, feeling, memory, volition and other psychological processes. This aspect has been variously called a soul, mind or consciousness. The fusion, integration or communication of mind and body, or soul and body, is an age-old puzzle that lies also at the heart of neuroscience.

The idea of a soul animating the body is thousands of years old. Some thinkers have located it in a specific organ and others have had it distributed through the whole body, as here in this 17th-century image of a man's soul by Johannes Amos Comenius.

Body and soul

Whether ancient thinkers regarded the heart or the mind as being in control of the body, they had to have a concept of what was doing the controlling. The notion that humans comprise both corporeal and non-corporeal elements – a body and a soul or mind – has existed since the earliest written records and probably predates writing. It lies behind religious concepts of an afterlife, in all their many forms, and provides a means of explaining life and mental activity.

Soul, psyche, spirit or mind

It is evident that something distinguishes living things from non-living things. Even the terms we use for this – animate and inanimate – enshrine the notion that there is a soul (*anima*) which 'animates' living things. A tiger is living, but a stone is not. A plant is living, but a door-knob is not. We have little difficulty, in general, deciding whether or not something is alive. The naming of some life force, mind, soul or animating energy helps to make and explain the distinction between living and dead; the loss of the force explains the transition at death to inanimate matter.

Homeric poetry of the 9th century BC or earlier considered the soul, or psyche, to be the 'life breath' which distinguished the living from the dead and departed on death to dwell in the underworld. Only humans were shown with souls, and only human souls were found in the underworld.

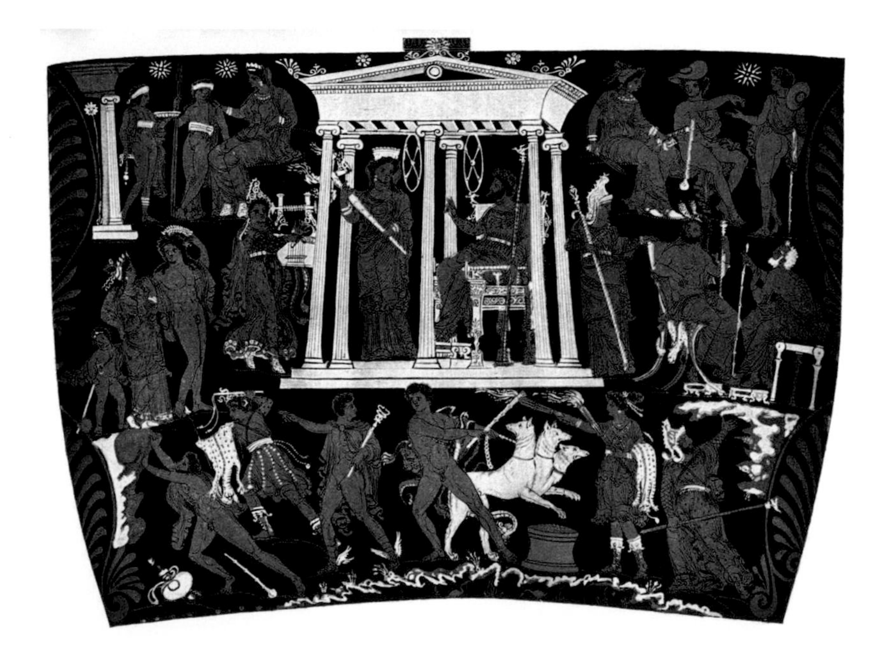

This depiction of the underworld on a Greek vase shows the shades of the dead looking indistinguishable from living characters.

Aristotle rejected the notion that the body and soul were two entities that could be spoken of separately in a meaningful way. For him, the psyche was simply the abilities or functions of a being. To talk of body and soul as Christian tradition does, or even as Aristotle's own tutor Plato did, was meaningless, for the psyche existed only so long as the organism was able to perform its functions and abilities. It is as foolish to ask 'whether the wax and its shape are one' as whether the body and the psyche are a single unit; they are integral, 'just as the pupil and sight make up the eye'. This put Aristotle in the position of locating all psychic function in the physical body.

THREE GRADES OF SOUL

In Aristotle's view, the capabilities of an organism are matched to its needs as a living being. Its ability to fulfil these needs defines the type of soul the organism has. Plants have simple needs and therefore simple capabilities, so they have the simplest of 'souls'. They can reproduce, nourish themselves and grow. Animals have more complex needs and abilities, as they also have powers of locomotion, perception and sensitivity. Finally, humans have the needs and capabilities of the nutritive (plant) and sensitive (animal) psyche, but also have rational powers.

The different types of psyche were thought to infuse and enliven the organism, enabling it to perform its functions while living.

The matter of souls

In the Greek model, then, a type of *pneuma* was considered a vital principle responsible for thought, feeling and the impulse to action. It was a physical substance, composed of a mixture of air and vapour from hot blood.

SOULS AND MAGNETS

By the 5th or 6th century BC in Greece, 'soul' was applied to other living things, as well as humans. Thales of Miletus even spoke of magnets being ensouled, presumably because they could initiate movement in susceptible matter (such as iron). Personal qualities were ascribed to the soul, too. We still retain this idea – even if we no longer believe in its literal truth, we use metaphorical phrases such as a 'kindly soul' to describe a compassionate person. The soul then becomes responsible for (or a repository of) characteristics and personality traits.

The dominant belief among the Ancient Greeks was that all matter below the heavens was composed of the four elements or roots – earth, water, air and fire – set out in the 5th century BC by Empedocles. The mix of these elements provided the characteristics of different types of matter. The soul was not made of a different type of matter from the body, just a more refined version with smaller particles. The soul was thought to be mostly air and fire, which are light and motile, whereas other matter included more of the heavy and sluggish elements – earth and water. If body and soul were made of essentially the same type of matter, this meant there was no particular problem in thinking of them as interacting. Aristotle's account of the *pneuma* being refined to different degrees was fully in accord with the idea that the soul was composed of ordinary matter (though Aristotle does, just once, suggest that it might have something akin to the special class of matter called 'aether' which he believed made up the celestial spheres).

The philosopher Epicurus (341–270BC) and the Stoics (starting in the 3rd century BC) considered the sympathy between body and psyche sufficient to prove that the psyche is the same as the body. As soul and body affect each other, and only material bodies can have an effect on other material bodies, the two had to be of the same nature.

Epicurus suggested that the psyche is made up of very fine matter, which is distributed throughout the physical body. Therefore, body and mind are an aggregate

Ancient Greek models of the cosmos had all corporeal matter made up of four elements: earth, water, air and fire. Aristotle added a fifth, incorporeal element, aether.

rather than the soul being located in a single region or organ. Acts of sensation and perception are possible because the mind is integrated into the body parts, so readily works in sympathy with them, a notion repeated by Nemesius in the 4th century AD.

> *'No incorporeal interacts with a body, and no body with an incorporeal, but one body interacts with another body. Now the soul interacts with the body when it is sick and being cut, and the body with the soul; thus when the soul feels shame and fear the body turns red and pale respectively. Therefore, the soul is a body.'*
>
> Nemesius, c.AD390

The Roman philosopher Lucretius (99–55BC) envisaged a tripartite division of body and soul, with the soul divided into a thinking part (*animus*) and a sensing part (*anima*). Again, he saw them as inextricably linked and, again, as made of the same type of matter, just of different gradations. He pointed out that a physical blow affects the senses and thought. Likewise, depression has an impact on the physical body and on sensations and thoughts – 'the nature of mind and soul is bodily'. Lucretius did not hold with the ancient view of a soul that survives death and goes to the underworld; in his opinion, the particles making up the mind, body and soul drift apart at death. All the matter is reused, but the unique person they once created has gone for good.

God gets in on the act

With the advent of Christianity, the business of the soul was hijacked by the Church. There have been both tripartite and bipartite models of the composite of body and soul. In the tripartite view, the soul, spirit and body are each separate; in the bipartite, there is only body and soul, spirit being simply another name for the soul. The soul was considered the immortal part that strives for communion with God. It was a fragment of the Holy Ghost, not made of corporeal matter, and was considered to survive after death and to be encumbered by the body.

The Church was not overly concerned with how the parts (soul, mind and body) communicated, but was largely clear that the urgings of the body generally put the soul in jeopardy unless the individual, either through reason or grace, managed to subdue the instincts of the body and act virtuously. The dominant paradigm was one of battle, or at best tension, between body and soul. The Church certainly did not consider the soul to be part of the body or to reside in a particular organ within it. Where a location for mental activity is mentioned, it is most frequently the heart, but this did not mean the heart was actually considered the tabernacle of the soul, as St Augustine (AD354–430) explained: 'In the Holy Scriptures . . . the term heart, a name for a part of the body, is applied to the soul in a metaphorical sense, while these philosophers maintain that it is the very organ that appears when the viscera are exposed.'

Augustine localized sensory and motor functions and memory in the brain, but

The fate of the soul after death has been disputed for millennia. In this 14th-century fresco, it will go to either Heaven or Hell.

retained reason and intellect as operations of the unlocalized soul. By the Middle Ages, *spiritus* had taken on an aspect of divinity which it did not have for Galen. For him, it was not in any way supernatural.

Several spirits

Galen's text came down to the European Middle Ages by a circuitous route. It was developed by Arab philosophers and medical writers before being reintroduced to southern Europe from the 12th century. Ibn Sina (see page 16) added his own views, describing the animal spirits undergoing change as they move from the first to the third cell, acquiring progressively more advanced capabilities. In the first cell they are capable of perception and imagination; in the second cell they add cognition; and in the third cell they are capable of memory. There was a clear change here from Galen in that it was not, in Ibn Sina's view, the ventricle but the spirit contained in it that had these abilities and functions. Neither did he think the spirit in the sensory or motor nerves to be the same as the spirit in any of the ventricles, but to represent an entirely different type of spirit.

Where is it?

If the body is animated by a soul, a reasonable question to ask is: Where is the soul located? It could suffuse the entire body, but that would present problems in the case of injury. What if a limb were lopped off? What would become of the bit of soul that was in it? In 1533, the Catholic Church ordered an autopsy after the death of conjoined twins Joana and Melchiora Ballestero to determine whether they had separate souls or shared a single one. The autopsy found two hearts and concluded that the twins therefore each had a soul, suggesting the Church believed that the soul was located in the heart.

It was almost as though the entire brain/heart debate had never been resolved: although control of the muscles and senses had been conclusively located in the brain, the soul – or perhaps even thought – could still be in the heart.

Two in one?

The fragmentation of spirit didn't really help to explain how the different senses or cognitive abilities worked. In the 17th century, the Cistercian monk Eustachius a Sancto Paulo (1573–1640) suggested combining all the spirits into one. He called this imagination and located it in the middle cell, leaving the first for sensory perception and the third for motor control.

> *'I suppose the human being to be nothing other than a statue or machine made of earth.'*
> René Descartes, *Treatise of Man*, 1662 (published posthumously)

The mechanical body

At the same time, the French philosopher René Descartes (1596–1650) promoted a mechanistic model of the human body. He had been inspired by seeing automata in the gardens of the Palace of Versailles, near Paris. The automata were moved by the pressure of fluids flowing through pipes, fed from underground. Seeing that simple hydraulic pressure could be harnessed to produce movement, and knowing also of clockwork mechanisms, Descartes speculated that the human body could also follow physical laws. He believed a purely mechanical explanation could be found for: 'the digestion of food, the beating of the heart and arteries, the nourishment and growth of the limbs, respiration, waking and sleeping, the reception by the external sense organs of light, sounds, smells, tastes, heat and other such qualities.'

Descartes went further, saying that the nutritive and sensory/locomotory aspects of

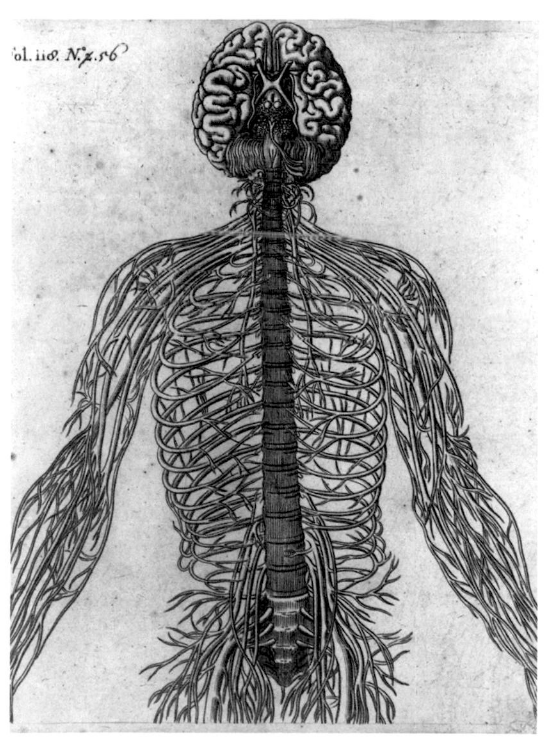

The brain and nervous system from Descartes' De homine.

the classical conception of the soul were not needed; these could be handled entirely by the mechanics of the physical body.

All types of physical or mental activity that are also carried out by other animals Descartes considered to be entirely mechanistic and to require no kind of soul or spirit. A dog or bee can perceive colour, hear sounds and respond to a pain stimulus, so clearly the human rational, immortal soul takes no part in those activities. But what of imagination, passion, conscience and those other faculties considered unique to humans? While the conceit of the body as a machine worked well enough for physical processes such as the flow of blood or the inspiration and expiration of air, and even perception, it could not account for the thinking, conscious 'I'. Descartes still required a rational soul to be hitched to the mechanistic body.

Following the same path as Eustachius, he proposed a single entity to encompass all types of purely mental activity including thought, imagination, will, reason and consciousness. He called this part the *res cogitans* and it equated closely enough to consciousness. He had effectively solved the problem of retaining the specialness of humankind, but he raised another problem. How could the soul, which he took to be rarefied and to have no physical substance, interact with the body? How can something that is not material move and change matter? It is undoubtedly the case that the soul or mind (for want of a better word) interacts with the physical body: we decide to move an arm and it moves; we shed tears when sad or hurt. In countless ways our minds and bodies interact and impact on each other.

The point of contact

Descartes needed a point of contact between body and soul; he chose the pineal gland, a small structure buried deep within the brain. It was a significant departure from the ventricular model in that he had chosen a solid physical structure, not a liquid-and-spirit-filled gap, as the locus of mental activity – though he did believe, wrongly, that the pineal gland is located in one of the ventricles. In having the pineal gland suspended in the fluid-filled ventricle, he could give a mechanical explanation for the interaction of body and *res cogitans*: 'The slightest movements on the part of this gland may alter very greatly the course of these spirits [flowing between the ventricles], and conversely any change, however slight, taking place in the course of the spirits does much to change the movement of the gland.' (1649)

He reasoned that as the pineal gland does not have two lobes and is located between the two hemispheres, it is well placed to take input from both hemispheres, so from both sides of the body, and fuse them into a single impression for the soul to comprehend. The pineal gland is also very small. Its near approximation to having no physical extension might also have appealed to him as the home of the *res cogitans*, which itself occupies no physical space.

Although Descartes had satisfied himself with finding a physical location for the soul, few others were persuaded. Through the 17th and 18th centuries, anatomists continued to try to locate the source of the 'animal spirit' and with it the seat of the soul or of consciousness. The brain scientist Thomas Willis declared the brain to be the

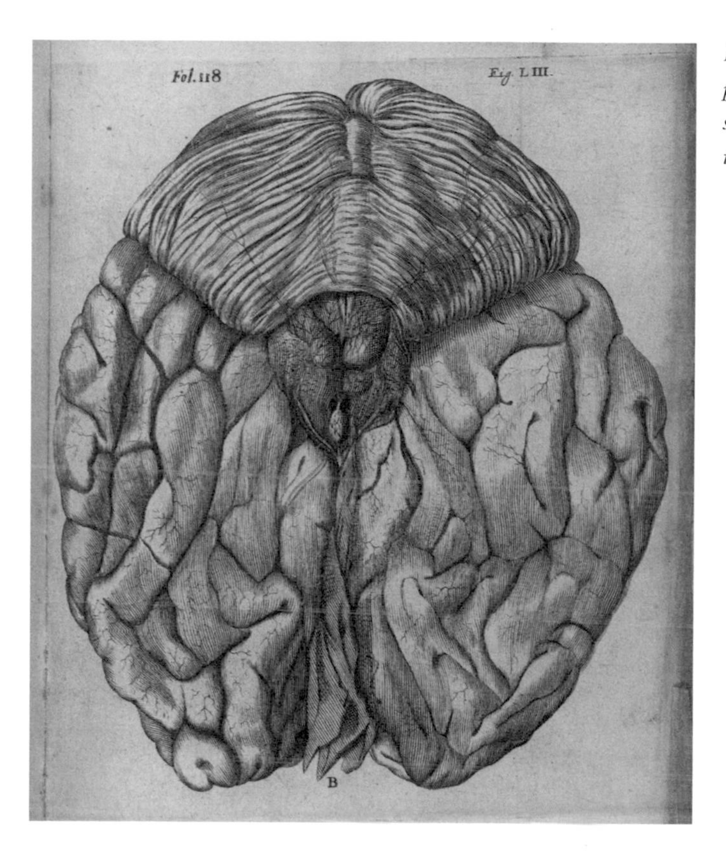

Descartes' diagram of a posterior view of the brain shows the pineal gland in the centre.

'living and breathing Chapel of the Deity'. He considered it likely that the cerebellum was the crucial organ, since damage to it is always fatal. Some others considered the corpus callosum to be the source of the animal spirits and yet others picked different locations, but the ventricles had fallen out of favour.

Consensus was never achieved: the difficulty of reconciling the body–soul duality remained intractable. Neuroscience still does not have a definitive answer: there are neuroscientists who believe in the existence of some metaphysical entity, and neuroscientists who believe the entire construct is material.

Unpicking the mechanism

The materialist position removes the need to explain how interaction is achieved. It offers the prospect of eventually discerning exactly how all aspects of the body work, including the brain engaging in its most insubstantial activities. Danish physician Nicolaus Steno put it succinctly: 'Since the brain is a machine, we have no reason to hope to discover its design through means any different from those used for discovering the design of other machines. The only thing to do is what we would do with other machines, taking apart its components piece by piece and considering what they do, separately and together.'

The dominant view retained some role for the immaterial mind/soul, but still left plenty of scope for taking apart the bodily mechanism, piece by piece.

Beginning in the 16th century, scientific exploration of the anatomy and physiology of the brain, nerves and spinal cord unravelled an increasing number of the mechanisms which enable our physical and mental interaction with the world around us.

The shrinking domain of the soul

As understanding about the physiology of the nervous system grew, the space left for an immaterial *res cogitans* to be effective was diminished. An English physician, Thomas Laycock, carried out extended studies of 'hysteria', a catch-all term for nervous behaviour (see box opposite). Reporting his findings in 1839, he included the statement that 'the cranial ganglia, although the seat of consciousness and will, are subject to the same laws which govern the other ganglia.' (A ganglion is a group of nerve cells, or the bodies of nerve cells; there are spinal ganglia along the spinal cord and cranial ganglia within the brain.) The suggestion that perhaps even some mental functions might be amenable to physiological explanation was astonishing. The scene was set for the inevitable extension of this principle to the whole of the mind/brain organ.

John Hughlings Jackson (see box on opposite page), perhaps most famous for his work on epilepsy, took the final step – denying that any kind of soul or metaphysical entity was required to make

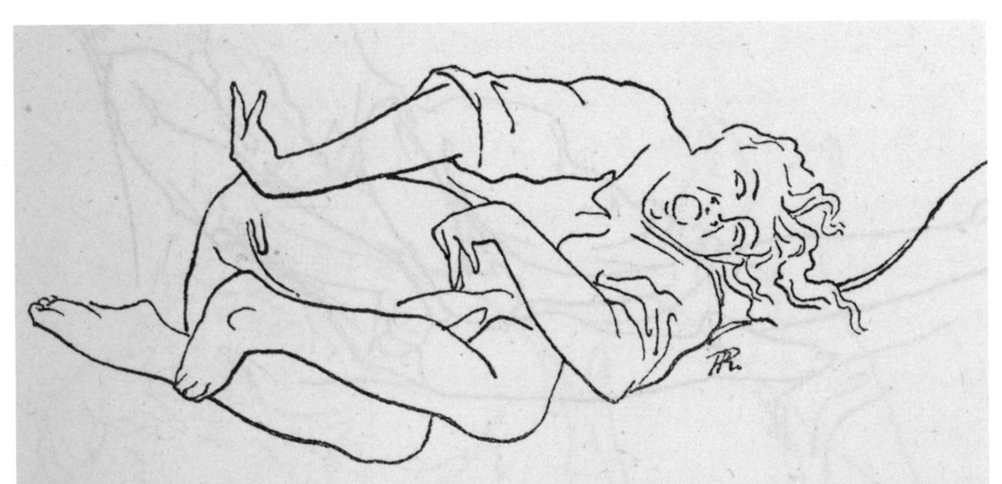

Epilepsy, depicted here in 1881, has provided fruitful research opportunities for neuroscience.

the human mind and body work. He saw no reason why the brain should be different from any other physical organ in principle, apart from in complexity. He drew on the newly discovered laws of the conservation of energy and the idea that the nervous system comprises a set of discrete organs working together. Building on these, his preferred model was of an entirely sensorimotor nervous system, including the higher aspects of mental functioning.

Three models

Hughlings Jackson recognized three possible solutions to the mind–body problem. The first was Cartesian dualism with the mind, in Jackson's words, 'act[ing] through the nervous system through an immaterial agency'. The second, materialist, model made the mind and body identical; 'the activities of the highest centres and mental states are one and the same thing', so the mind is entirely physical. The third model saw brain and

THE SPIRIT GOES AWOL

It is difficult to study the function of the brain, or of the mind/spirit, in the context of its normal functioning. Throughout the history of neuroscience researchers have drawn extensively on evidence drawn from malfunction and difference. Brain lesions (damage caused by disease or accident), epilepsy, mental illness of various types, neurological diseases such as Alzheimer's and special states such as hypnosis all provide insights by contrast. Epilepsy has proven a particularly rich area for research. Hysteria, which was once associated with epilepsy, is no longer considered a discrete diagnosis. It is now considered a type of somatizing: this is when the body converts psychological distress or stress into physical symptoms that can include erratic behaviour, loss of consciousness, seizures, selective amnesia, elective mutism and many others.

JOHN HUGHLINGS JACKSON (1835–1911)

Hughlings Jackson trained in medicine and later worked in London at the London Hospital and at the National Hospital for Paralysis and Epilepsy. His most famous work was with epilepsy patients, documenting in minute detail the progress of seizures and identifying different forms.

He was interested in philosophy as well as physiology, and proclaimed his aim to discover the workings of the entire nervous system. An agnostic who disregarded the metaphysical, he believed such an explanation could encompass all that is involved in being human: 'A man, physically regarded, is a sensorimotor mechanism. . . If the doctrine of evolution be true, all nervous centres must be of sensory-motor constitution.' (1884)

Sick supplicants hope for a miracle cure from a reliquary. The man at top left suffers from epilepsy or mental illness, signified by the bat-like demons above his head.

mind as different, but acting in parallel: there was no causal interaction between them, but they followed the same path. A similar idea, suggested by German philosopher Gottfried Leibniz in the 17th century, used the analogy of two clocks set in motion at the same point, both telling the same time even though there was no link between them. Hughlings Jackson called this the Doctrine of Concomitance. While he rejected the metaphysical, Hughlings Jackson was enthusiastic about the Doctrine of Concomitance because it allowed neurology free rein within the body, without having to deny (or heed) an immaterial *res cogitans*.

Prised apart

In adopting the Doctrine of Concomitance, Hughlings Jackson effectively freed neurology to develop as a clinical discipline in its own right. It could deal entirely with patients' sensorimotor signs and symptoms and disregard any mental factors. Although Hughlings Jackson, like later neuroscientists, was aware that state of mind and emotionally loaded events could have an impact on physical health, he believed that they were not to be taken into account in assessing and treating the sensorimotor aspects of neurological illness. This had the additional impact of forcing neurology and psychology apart. Psychology would deal with the immaterial mind, the relation of the self to the world, while neurology would deal with the physical manifestations of disease and distress. Neuroscience has followed the path he cleared, though rather than ignoring the concerns of psychology it attempts to explain or elucidate them by reference to physical processes in the brain.

Although Hughlings Jackson had separated the mental and the physical, he also rejected the idea of the unconscious mind. He felt that if there was something the mind did below the level of human consciousness, it should be apparent in unconscious patients, but nothing is apparently happening when a patient is unconscious (he could not use modern technology to test brain activity, of course). Ultimately, he just wanted to be rid of the complexity of the mind: 'As an evolutionist I am not concerned with this question [of the mind], and for medical purposes I do not care about it.' (1888)

CROSSING THE LINE

Hughlings Jackson closed his eyes to the impact of the mental and emotional on the body, but these factors are becoming increasingly important today. Neurologists often see patients with genuine physical symptoms but for whom no somatic (body-related) cause can be found. The idea that their illness is 'all in their mind' or psychosomatic is distressing to the patients – it implies they are making it up. But somatized pain is not made up: the pain is as real an experience as pain with a physical cause. As we shall see in Chapter 7, all pain is produced in the brain.

Knowing the emotional distress that is being somatized is useful to the clinician. For one thing, it can put a stop to invasive treatments and explorations that will never work because the problem is not in the part of the body that shows the symptoms. For another, it gives a chance of approaching effective treatment by dealing with the situation that is causing the patient to produce physical symptoms. The difficulty of confirming with confidence that a symptom is somatized distress rather than an undiagnosed condition is considerable, and relies largely on ruling out other options. Somatization is an example of the mind overstepping the mark, as it were, in its influence on the physical body.

SLYNESS
PRIDE
SLEEPINESS
BUMPS
Doctrina Particularum
Self Knowledge
Commentaria Critica
Belle Brain
Belle Brain
Aristotle
Sold by
ROYAL PATENT
PHRENOLOGICAL
HATS
ADAPTED TO EVERY
PROTUBERANGE
of
FACULTY OR ORGAN
DISCOVERED
ABSTRACTION
SCOTT
TREMAINE
Pollard
De Ville

CHAPTER 3

The localization of **BRAIN FUNCTION**

'Among the various parts of an animated body which are subject to Anatomical disquisition, none is presumed to be easier or better known than the Brain; yet in the mean time, there is none less . . . perfectly understood.'

Thomas Willis, 1644

The brain looks fairly homogeneous to casual observation, yet it was already, by the 16th century, credited with a range of abilities. The question naturally arose as to whether it has specialized areas (even discrete organs) or whether all these functions are jumbled together. Our best answer, after about 500 years' work, is that it's a bit of both.

An early method of trying to assign different functions to different parts of the brain was phrenology, ridiculed by the medical profession but enthusiastically adopted by many others.

One thing or many?

Even the earliest models of the brain made some attempt at localizing functions in different regions, but these were based more on thinking about how the brain *should* work than looking at how it *does* work. Galen's assumption that sensory input and motor control are handled at the front of the brain, and the mental faculties in the ventricles, set a pattern of general localization which was developed and refined in coming centuries. Finally, in the 17th century, ideas about localization began to be based on anatomical study rather than philosophy and tradition.

Plato used a chariot allegory to describe the human soul. The charioteer represents intellect and reason; the horses represent emotion and desire. The charioteer is tasked with steering the horses on a steady course towards enlightenment.

At the end of the 15th century, human and animal dissection began again after a hiatus of around 1,800 years. Not only did anatomists examine their subjects with care, they also recorded what they saw in illustrated texts. From the time of Vesalius in the 16th century, anatomists found, or at last acknowledged, that many structures of the body (including the brain) do not match the accounts given by Galen.

Although more accurate accounts of the brain's structure emerged as the anatomists set to work, their understanding of what the brain does or how it does it remained unclear. Much of the brain looks relatively featureless. The mass of the cortex, dismissed as 'rind' early on, does not have obvious fine structures like the small projections on the lining of the stomach (villi) or the air sacs of the lungs (alveoli).

Towards neurology

The first person to investigate the anatomy of the brain thoroughly was the English physician Thomas Willis (1621–75). A professor of 'natural philosophy' (science) at Oxford, Willis produced an influential monograph called *Cerebri Anatome* (*The Anatomy of the Brain*) in 1664. In this he gave a detailed account of the structure of the brain and coined the term 'neurology'. Willis did not set out with the agenda that a modern anatomist would have; he stated that he wanted to 'unlock the secret places of Man's Mind and look into the living and breathing Chapel of the Deity'.

Dissection lay at the root of Willis's work. He directed dissections that were carried out by his assistant, the physician Richard Lower, in the back rooms of inns

> *'Men ought to know that from nothing else but the brain come joys, delights, laughter and sports, and sorrows, griefs, despondency, and lamentations. And by this, in an especial manner, we acquire wisdom and knowledge, and see and hear, and know what are foul and what are fair, what are bad and what are good, what are sweet, and what unsavoury; some we discriminate by habit, and some we perceive by their utility. By this we distinguish objects of relish and disrelish, according to the seasons; and the same things do not always please us. And by the same organ we become mad and delirious, and fears and terrors assail us, some by night, and some by day, and dreams and untimely wanderings, and cares that are not suitable, and ignorance of present circumstances, desuetude [inertia], and unskilfulness.'*
>
> Hippocrates, *On the Sacred Disease*, *c.*400BC

and private houses. Willis examined the structures exposed using a magnifying glass or microscope, and they were drawn by Christopher Wren (more famous as the architect of the new St Paul's Cathedral). He injected dye into the blood vessels of the brain to trace their path, and did considerable work on the circulation of the brain.

Willis tried to work out the functions of the different areas he saw and described. He proposed that the gyri (the bulges on the cerebral cortex) control memory and the will, making him the first person to locate psychological functions in the cortex rather than the ventricles. He attributed the many convolutions of the human cortex to the greater psychological abilities of humans (than of other animals). Visual perception he attributed to the corpus callosum, a broad band of nerve fibres that joins the two hemispheres and is the largest mass of white matter in the brain. He seems to have envisaged it as rather like a screen on which images were projected for the rational soul to observe. He also attributed other types of sensation and movement to the corpus striatum. Involuntary movement and vital functions he assigned to the cerebellum, at the lower back of the brain.

Willis's suggestions regarding function were informed more by his ideas about the soul than by empirical evidence. He believed in three types of soul, roughly following the pattern of Aristotle and Plato. In addition to the sensitive soul and vital soul shared with animals, humans have an immortal soul capable of higher thought, will and judgement. The immortal soul had no material form but did, in Willis's view, act upon the brain. He had no explanation for how the two might interact. The material soul (shared with animals), he explained in some detail. Following Galen, he described how animal spirits present in the brain and nerves are refined from the vital spirits which circulate in the blood. The animal spirits, he said, are generated in the cortex and cerebellum (not in the third cell, as the traditional model proposed) and stored in the brain. They travel along the nerves to the sense organs and muscles as needed. He described something like a reflex arc

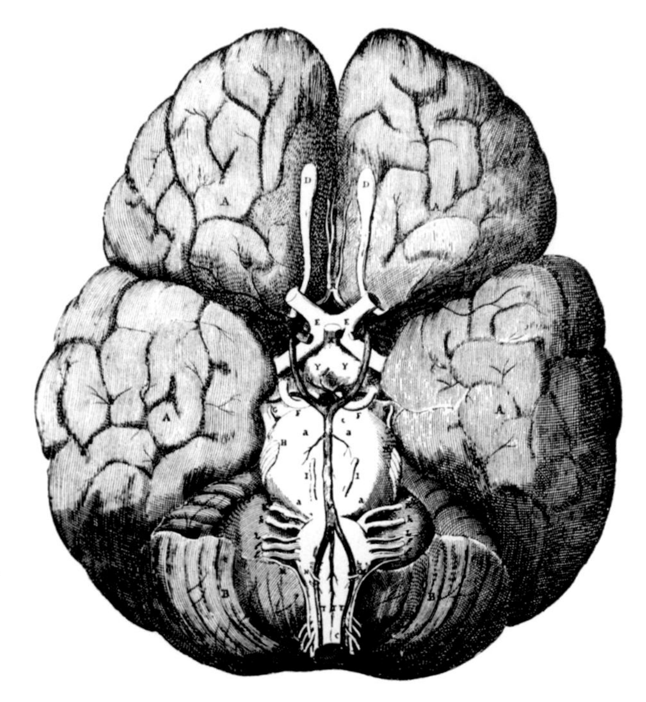

Willis did not follow the standard procedure for the time. Instead of carrying out the dissection in situ, he had the brain removed first and approached it from below, as shown here in Christopher Wren's drawing from Cerebri Anatome, *1664.*

whereby sensory perceptions are processed, then the cortex initiates the flow of animal spirits to the muscles. This explains many types of action in a way that is available to both humans and animals, so animals (which he considered lack volition) 'move themselves or their members, only as they are excited from the impulse of the external object, and so sensation preceding motion, is in some manner the cause of it'. Humans have an additional way of responding. The corpus callosum, on perceiving images projected on it, can initiate willed actions.

Willis's description of the structures of the brain was detailed and meticulous, and the first to give the current numbering of the cranial nerves, which emerge from the brain and connect to parts of the head and neck. He distinguished between white and grey matter, making white matter responsible for the generation of animal spirits and grey matter responsible for their operation and distribution.

His book effectively shifted the locus of mental activity from the ventricles to the cortex, changing the direction of investigations of the brain. Even so, it did very little to reveal the workings of the brain and his broad localizations were not based on empirical evidence. Nicolaus Steno lamented the state of ignorance about the brain:

'We need only view a Dissection of the large Mass, the Brain, to have ground to bewail our Ignorance. On the very Surface you see varieties which deserve your admiration: but when you look into its inner Substance you are utterly in the dark,

ANNE GREEN, BACK FROM THE DEAD

Thomas Willis and his mentor William Petty often worked together on dissections, which they carried out in Petty's home. On one such occasion, they got more than they bargained for.

Petty was allowed to claim for dissection the body of any criminal executed within 21 miles of Oxford. On 14 December 1650, the pair prepared to anatomize the body of Anne Green, a scullery maid who had been raped and subsequently hanged for the infanticide of her newborn baby (the child was later discovered to have been stillborn). Green was hanged at Oxford, left suspended for half-an-hour, then removed to a coffin and to Petty's house. But when Willis and Petty opened the coffin and prepared to dissect the body, she made a strange sound and began to breathe. The two men revived her with hot cordial, tickled her throat with a feather to prompt coughing, rubbed her arms and legs and then put her to bed with another woman to warm her. Within 12 hours she could speak and in a month she was fully recovered. She was granted a free pardon and went on to marry and have three more children.

being able to say nothing more than that there are two Substances, one greyish and the other white.'

Localizing brain functions

Despite good intentions, locating the areas of the brain responsible for different activities proved difficult and little progress was made before the 18th century.

St Luke operating on a man's head.

Insight from injury

As Steno had observed, the brain presents a puzzle. It cannot be seen to be doing anything even if the head is opened up. But head injuries can have specific adverse effects, and these gave some indication as to what might be going on inside the brain.

In 1710, François Pourfour du Petit (1667–1741), a French military surgeon, treated a patient with a brain abscess. The man suffered from paralysis on the opposite side of his body to that of the abscess, leading Petit to conclude that the animal spirits traverse from one side of the brain to the other in the tracts which cross in the medullary pyramids (paired structures at the top of the brain stem, just below the pons). He demonstrated that he could produce translateral paralysis by severing the connection to one or other of the pyramids in dogs. In 1727, he went further, tracing the nerves that cross in the pyramids to their origins in the cerebral cortex. With this, he established for the first time the existence of the motor cortex.

This is a most basic lateralization of function, finding that one side of the brain controls and takes input from the other side of the body, but it is a very important discovery. Yet it was the only discovery to have much impact for some time.

Looking ahead

Emanuel Swedenborg (1688–1772) seems an unlikely candidate for a prescient neuroscientist. After studying theology and becoming interested in the beliefs of a dissenting Lutheran sect, he settled to working in natural science and inventions.

Emanuel Swedenborg gave up his research into the soul after a divine calling.

Among his proposals were one for a flying machine and another for a submarine.

Swedenborg's twin interests in natural science and religion led him to try to investigate the biology of the soul. He believed the soul to be connected to the body and based on material substances – therefore susceptible to study. Beginning in the 1730s, he undertook extensive research into the structure and function of the brain and nervous system and thought deeply and with originality, anticipating many later discoveries. His aim was to locate the soul, and he formulated an ambitious plan to publish a 17-volume work on its anatomy. He left his job in 1743 to gather material for his book, but the following year experienced a vision in which, he claimed, Christ said he had chosen to reveal to him the true meaning of the Bible. Not surprisingly, he had to give up his previous project to undertake this demanding divine commission. And demanding it was – starting from the Hebrew texts, Swedenborg was tasked with finding the spiritual meaning of every verse of the Bible.

Re-evaluating the cortex

At the time, the predominant view was that the cortex was unimportant in terms of brain function, its sole job being to deliver blood vessels to deeper parts of the brain where all the real work went on. The view that it did little of use was reinforced by the work of Swiss physiologist Albrecht von Haller (1708–77), the leading authority on the brain. He tested the 'irritability' (sensitivity) of various body tissues and found the cortex to be completely unresponsive. In experiments with dogs, he stimulated the cortex with a scalpel, with corrosive substances and with anything else he thought might cause pain, but the dogs remained blissfully unaware. Only when Haller plunged his instruments deeply into the brain did the dog howl and struggle. His conclusion was that the cortex is indeed just a rind with no sensory or motor function and not involved in higher mental functions.

Swedenborg read the literature on the brain concerning its physical structure, including experimental results and observation, then interpreted the data afresh and came to completely different conclusions from his sources. His main finding was that the cortex is the centre for both receiving sensory information and initiating willed action: 'the cortical substance . . . imparts life, that is sensation,

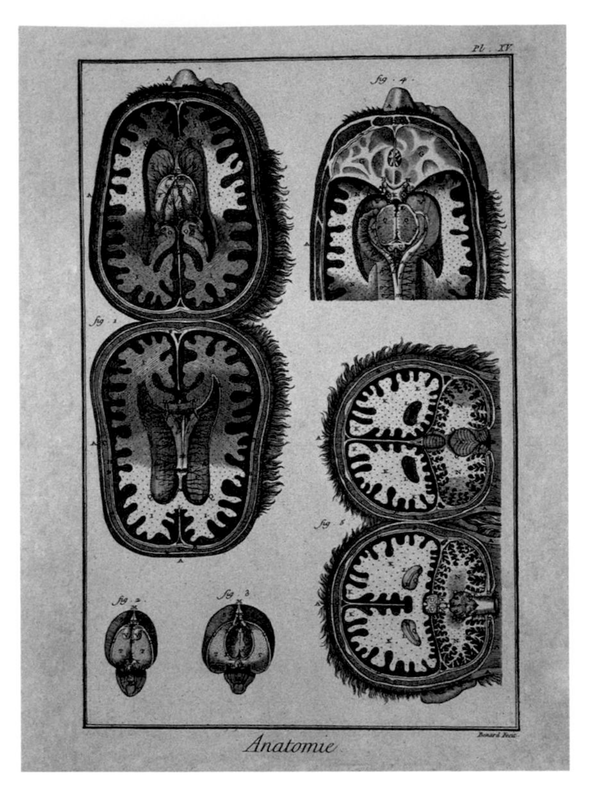

perception, understanding and will; and it imparts motion, that is the power of acting in agreement with will and with nature.'

Italian biologist Marcello Malpighi (1628–94) was the first person to use a microscope to examine the cortex. He reported that it was made up of many small glands, or 'globules', with attached fibres (see page 77). The globules were later shown to be artefacts produced by Malpighi's microscope and the way he prepared his samples. Swedenborg concentrated on the fibres Malpighi had seen, suggesting that they might be connecting independent units (the globules) which acted as 'cerebellula', or

An engraving of sections of the brain, late 18th century.

A CONFUSED ACCOUNT

Domenico Mistichelli, professor of medicine at the University of Pisa, had noted the crossing of nerves and the contralateral effect of brain lesions the year before Pourfour du Petit, but his account was much less enlightening:

'The medulla oblongata externally is interwoven with fibres that have the closest resemblance to a woman's plaited tresses . . . many nerves that spread out on one side have their roots on the other; so, for example, those that extend to the right arm, through such plaiting, can readily have their roots in the left fibres of the meninges. The same may be understood of those on the left proceeding from the right. . . . Therefore the supposition is clear that, if on the right side . . . through oppressive humours, or through convulsions, strangulation, or some other defect, the transit of the animal liquid [spirit] through very small interstices is impeded, it will soon happen that the arm or leg or other left part, with which those nerve filaments are in agreement, will remain either convulsed or paralyzed, or deprived of sensation and motion, because the nerves of those parts do not receive the necessary supply of spirit from the opposed part that has been injured.'

> *'I have pursued [the study of brain] anatomy solely for the purpose of discovering the soul. If I shall have furnished anything of use to the anatomie or medical world it will be gratifying, but still more so if I shall have thrown any light upon the discovery of the soul.'*
>
> Emanuel Swedenborg

mini-brains. In identifying these as discrete elements that work in conjunction with one another, he brilliantly anticipated the doctrine of the neuron (brain cell) that emerged more than 100 years later in the 1890s (see page 99). The fibres run through the cortex to the white matter, through the medulla, down the spinal cord and then to the parts of the body by means of the peripheral nerves. These, he proposed, were the conduits of sensation and action.

Swedenborg was certain that sensations end up in the cerebral cortex, as that is where the nerve fibres had their origin or end. He was not clear about whether different sensations are localized in different areas of the cortex, but did propose that motor control is localized, saying (correctly) that control of the foot is located in the dorsal (rear) cortex and of the face and head in the ventral (frontal) cortex. This insight did not resurface until 1870.

On the localization of mental functions, Swedenborg noted that injury to the front of the cerebrum was more likely to damage the 'internal senses – imagination, memory, thought' – saying that 'the very will is blunted'. However, he stated that damage to the back of the brain did not have this result. Of the pituitary gland, he wrote that it is the 'crown of the whole chymical laboratory of the brain' – a view which reappeared in the 20th century. Swedenborg said the *corpus callosum* allowed the two hemispheres of the

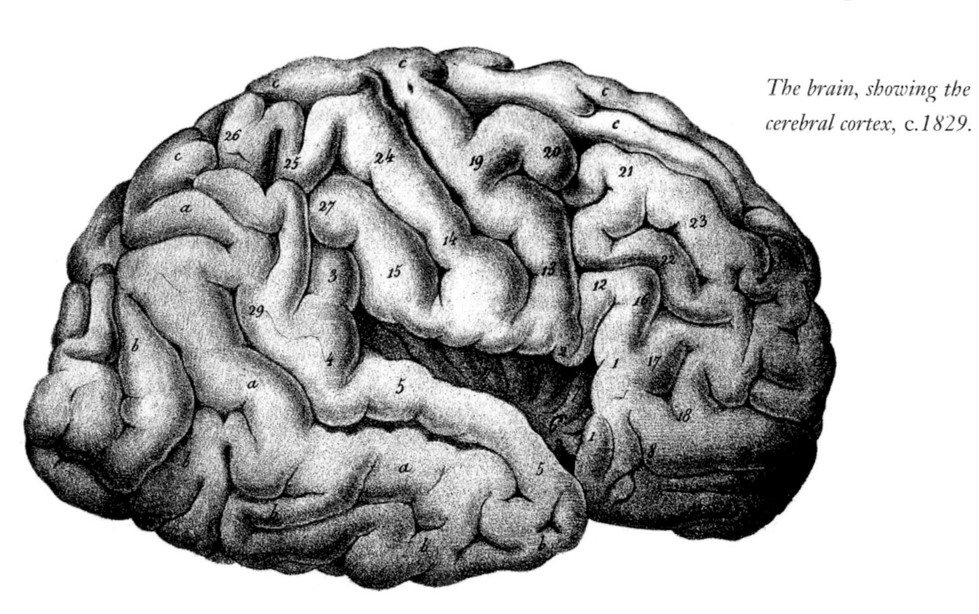

The brain, showing the cerebral cortex, c.*1829.*

brain to communicate with each other (which it does) and that the *corpus striatum* took over the function of motor control when a movement became 'second Nature'.

Marcello Malpighi is best known for discovering the blood capillaries.

A voice in the wilderness

For all these astonishing insights, Swedenborg had no influence on the development of neuroscience, probably because his physiology was presented in the context of his search for the soul, so was not noticed by scientists. In the 1880s, after Gustav Fritsch and Eduard Hitzig had discovered the motor cortex (see page 57), there was a flurry of interest in Swedenborg as people realized he had already said much that was newly 'discovered', but he vanished back into obscurity again soon afterwards.

The slow path to discovery

Back in the mainstream, progress limped along. Inevitably, it was laboratory animals that bore the brunt of early investigation into what parts of the brain might be doing. In 1760, the French physiologist Antoine Charles de Lorry removed the cerebellum and cerebrum from dogs and reported that they continued to breathe for 15 minutes. He concluded that the medulla, previously considered just an extension of the spinal cord, must be responsible for vital functions.

In 1806, Julien-Jean-César Legallois carried out experiments to find out exactly which part of the medulla housed the respiratory centre. He removed the cerebellum from young rabbits, then removed slice after slice of the midbrain and medulla. He found that when he severed the medulla at the level of the eighth cranial nerve, the rabbits stopped breathing, so

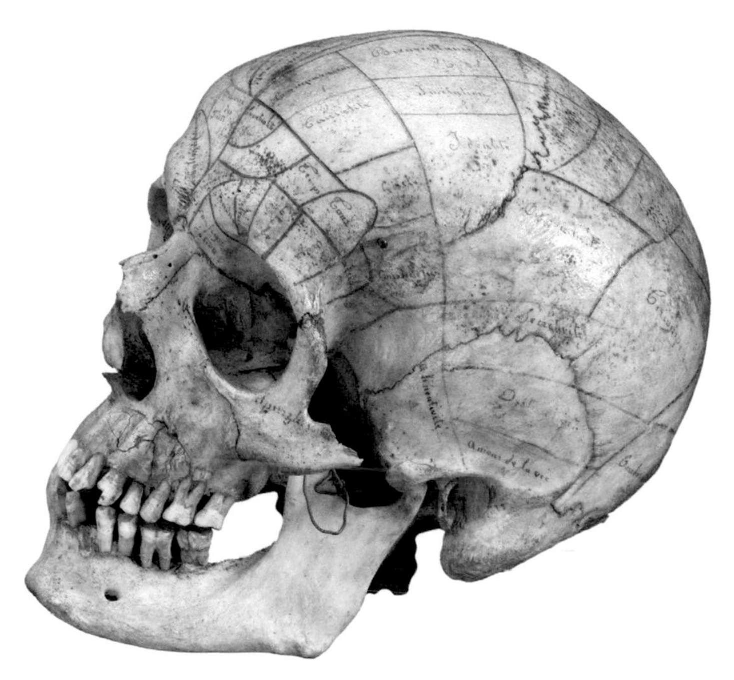

A skull marked with the areas that phrenologists identified with various character attributes.

he located the centre of respiration there. Marie-Jean-Pierre Flourens located it more precisely in 1851, reporting it to be no larger than the head of a pin (in rabbits). Legallois' finding was the first widely accepted evidence that functions really were localized within the brain. But, almost immediately, the project of charting brain localization was diverted by a maverick attempt to find the source of psychological functions and attributes of personality.

Lumps and bumps

The character most commonly associated with early ideas of brain localization is Franz Gall (1758–1828). A physician and anatomist, he worked in Vienna in the late 18th century but was forced to move to Paris after the Austrian government, under pressure from the Church, clamped down on his popular demonstrations.

Gall is famous for developing the pseudo-science of phrenology, which attempts to discover aspects of character by examining the shape of the skull. It is based on Gall's theory that the cerebral cortex is divided into 27 separate 'organs', each one having its own responsibilities. The size of each organ correlates to how well developed or prominent the corresponding faculty is in the individual. According to Gall, as an organ grows it pushes against the skull, forming a bump on the outside which an expert can identify and measure. Gall called his study of skull-shape to determine character 'organology'. He studied a collection of skulls and skull casts to arrive at his method, trying to match highly developed traits with unusual skull bumps. The size of the 'organ of benevolence', for example, was thought to determine how kind someone would be.

Although phrenology has been debunked, it was important in establishing the idea of localization. Furthermore, the idea that even personality traits can be localized has resurfaced with modern brain-imaging.

The growth and death of phrenology

In 1800, Gall hired a physician, Johann Spurzheim, as his assistant. Spurzheim soon became fully engaged in the project; Gall saw him as his successor and cited him as co-author of his books. Gall and Spurzheim fell out in 1812 and Spurzheim began a separate career developing phrenology further, increasing the number of organs to 35 and making the system hierarchical. He became very successful, touring Europe and giving lectures and demonstrations.

Ironically, phrenology was brought to public attention and growing popularity by a damning condemnation in the *Edinburgh Review* in 1815, which called it 'a piece of thorough quackery from beginning to end'. Spurzheim answered the criticisms in the article and won converts in Edinburgh. George Combe, a lawyer, saw the article and at first mocked phrenology but later became a convert and vocal proponent. Combe was criticized for being both a materialist and an atheist. In one book, *Constitution of Man*, he wrote that: 'Mental qualities are determined by the size, form and constitution of the brain; and these are transmitted by hereditary descent.' It was a view that was controversial and modern, taking on board the proto-evolutionary ideas current among progressive natural scientists.

Phrenology enjoyed so much popular success that some employers introduced phrenological character analysis as part of their selection process. Phrenology

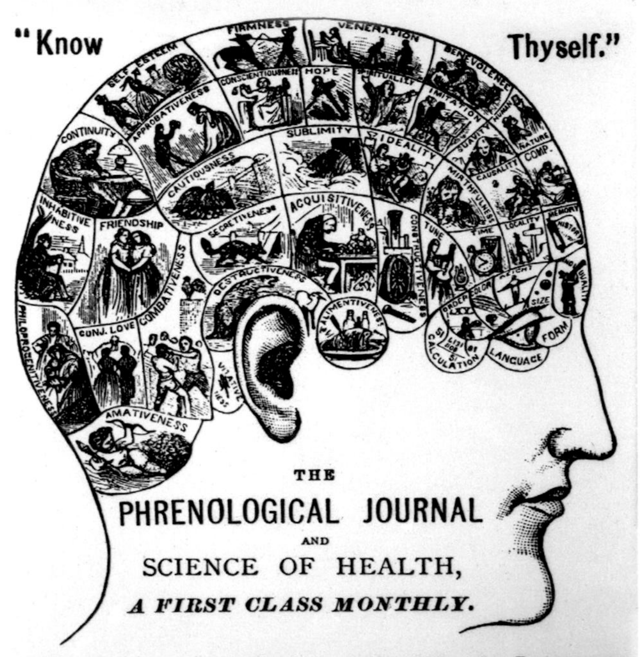

A phrenological map shows different aspects of personality in different areas of the brain.

FRANZ GALL (1758–1828)

Born in Baden, now part of Germany, Gall went to medical school in Strasbourg and in Vienna, Austria. He took a job in the lunatic asylum in Vienna and studied mentally ill patients. There he developed his ideas into a full account of how the size of different areas of the brain determined personality and could be 'read' from the shape of the skull. Gall claimed to have had the idea that lay behind phrenology when he was nine years old. He had noticed that a school friend who had bulging eyes had a better memory for words than Gall had himself. He noticed a similar trait among other students with a special facility with words and later decided that the area of the brain that worked with speech must be in the frontal lobes. He believed that a well-developed verbal centre would push the eyes forwards, making them bulge.

Gall opened a private practice and held popular public lectures explaining his theories. His ideas were popular with the public but not with the authorities, and he had to move first to Germany and then to France. Opponents of the principles of 'organology' criticized it for being unscientific, immoral and anti-religious. It was enthusiastically adopted by European anthropologists of the 19th and early 20th centuries, as it seemed to be a way of 'proving' that Europeans were superior to other 'races', so excusing the atrocious behaviour of the colonizers towards the people they conquered or exploited. It was most popular in Britain and France, and later in North America.

Gall made significant discoveries besides the development of phrenology. He was the first person to determine that grey matter is functioning neural tissue and that the brain is folded so it can fit a lot of surface area into a relatively small volume. In addition, he proved that motor nerve fibres cross (decussate) as they leave the brain stem and enter the spinal cord at the medullary pyramids.

busts became widespread, and all kinds of people set up in business as phrenologists. Some simply felt the head with their hands, as Gall recommended, but others used calipers to measure the contours more accurately.

Enthusiasm for phrenology waned in the mid-19th century, but it took a long time to die – the British Phrenological Society was only finally disbanded in 1967. Two central tenets of phrenology, that some abilities are localized in the brain and that repeated use can grow certain parts of the brain, are now accepted by modern neuroscience, but the notion of reading character or brain structure from the bumps on the head has been shown to be unfounded.

Making sense of the motor cortex

By bringing the idea of localization into currency, Gall helped neuroscience to move in the right direction. Its progress, though, did not start from the psychological attributes that interested Gall, but from examining control of movement.

An early contributor was the French physiologist Marie-Jean-Pierre Flourens, who had identified the respiratory centre of the rabbit. A staunch opponent of Gall and phrenology, Flourens was convinced that all faculties were spread out through the brain.

Flourens' convictions were based in his experiments, most of which were carried out on animals. He found that if part of the cortex suffered a lesion (particularly

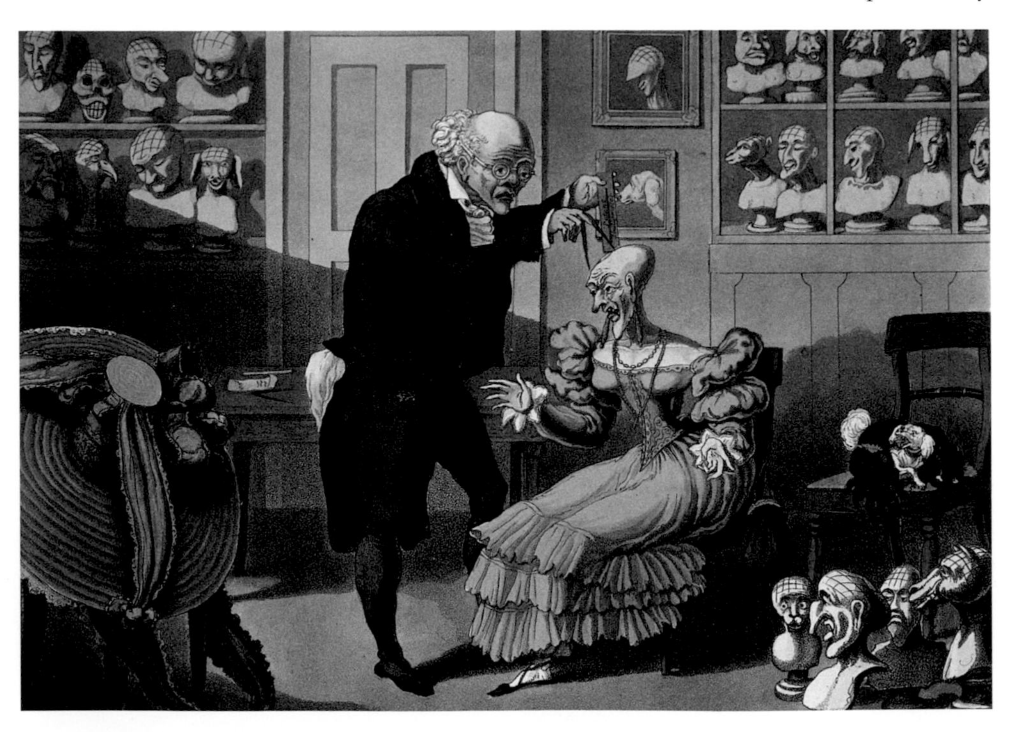

This woman has removed her wig to enable Gall to measure her head with his calipers.

in birds), recovery was either complete or non-existent: his experimental animals had all their faculties restored or none, suggesting there was no localization of faculty. He concluded that the cerebellum is responsible for coordinated movement and the medulla sustained vital functions, but the cortex could not be functionally divided.

Proved wrong by a dog

Flourens was wrong, though. In Germany in 1870, psychiatrist Eduard Hitzig and Gustav Fritsch, an anatomist, published the results of their experiments applying an electric current to the cortex of dogs. They had carried out their experiments on a dressing table in one of Hitzig's bedrooms.

Hitzig had developed a piece of equipment to administer therapeutic electric shocks to his patients. He found that if he applied current to the back of the patient's head, their eyes moved, and this was reliably reproducible. It prompted him to investigate further – and that's where the dogs came in. Hitzig and Fritsch administered a very light electric current to different parts of the dogs' cortices and recorded any corresponding movements.

They found they could isolate small areas that produced movement of the fore-paw, hind-paw, face and neck. In all cases, the movement was on the opposite side from the stimulus (so a stimulus on the left side of the brain produced corresponding movement on the right side of the body). They concluded that only part of the cortex was involved in motor responses, and it tended to be at the front of the brain. The centres were narrowly localized and responded to a very weak stimulus. Hitzig and Fritsch also found that if they removed or destroyed the area of the cortex which controlled a fore-paw, sensory responses from the paw were not affected – they had discovered that this area dealt only with motor control.

Fine details

Scots psychiatrist and neurologist David Ferrier (1843–1928) took their work further in the 1870s, experimenting mostly on monkeys. He used even smaller currents and produced very detailed maps of the motor cortex areas of the monkey brain. His work on fish, amphibians and birds failed to find any motor cortex response, supporting Flourens' findings with animals (which he had unfortunately and inaccurately extended to humans). He went on to discover the areas involved in smell and hearing.

Ferrier's detailed mapping of localized areas of the brain was soon used

Eduard Hitzig (centre with beard and glasses) and Gustav Fritsch (seated).

by neurosurgeons. Their findings in turn fed back into the mapping of the brain's functions.

Speech to the fore

In the mid-19th century, before Ferrier's work on the motor cortex, opinion was polarized between those who considered function to be localized, and those who followed Flourens and believed function to be distributed. Those in favour of localization were hobbled by the stigma of phrenology, never popular in the medical profession. It was against this background that a series of French medical practitioners struggled to demonstrate that at least one uniquely human attribute has a very precise locus in the brain.

Battling against phrenology

Evidence had been mounting in the middle decades of the 19th century that loss of speech often followed damage to the front part of the brain. French physician Jean-Baptiste Bouillaud claimed in 1825 that lesions in the frontal lobe caused the loss of articulate speech. Bouillaud had been a phrenology enthusiast early on, but moved away from it, though still promoted localization of brain function. He collected a large number of cases (he was the first brain scientist to work with a large dataset) and came to the conclusion that the centre of speech is at the front of the brain. He even demonstrated in 1827 that if he destroyed part of the brain of a dog between the anterior and mid sections, the dog lost the ability to bark.

The association of localization and phrenology made other professionals anxious. Critics pointed out that some people with damage to the front of the brain didn't have impaired speech. Bouillaud didn't locate the speech faculty very precisely, so the connection remained tenuous. In 1848, he famously challenged anyone to find a patient with similar loss of speech who did not have a lesion on the frontal lobe, offering a reward of 500 francs. The prize was finally awarded in 1865 to French anatomist and surgeon Alfred Velpeau for a patient whose frontal lobes had been destroyed or displaced by a cancerous tumour but who retained the ability to speak – by that time, however, the centre for speech had been conclusively located.

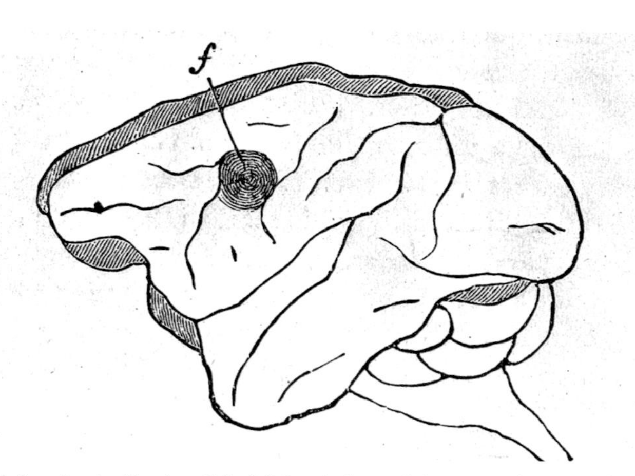

A drawing by Ferrier of the left hemisphere of the brain, showing a lesion that caused paralysis of the biceps.

An unfortunate suicide

In 1861, the French physician Ernest Auburtin described a patient who had shot himself in the head. The patient had blown away part of his skull but survived for several hours, during which time Auburtin conducted experiments on his exposed brain. (Medical ethics don't seem to have played much of a part in the early days of neurology.) Auburtin found that if he used a spatula to press on the front part of the brain while the patient was speaking, his speech stopped. When he relieved the pressure, the patient started talking again.

Jean-Baptiste Bouillaud was the first to offer evidence that the capacity for speech is located near the front of the brain.

Big and brainy?

Auburtin presented his findings at a meeting of the Société d'Anthropologie in Paris. He was of the view that if a single example of localized function could be confirmed then the debate would be put to sleep. Unfortunately, no one took very much notice of his suicide case. Another provocative presentation was also made at the same meeting. Anatomist Pierre Gratiolet described the very large skull of a North American Totonac Indian. It prompted a heated debate about whether intelligence could be inferred from brain size.

Dead men DO tell tales

The physician Paul Broca was among those interested in Auburtin's findings. In 1861, Broca accepted on to his surgical ward a patient near death, Louis Leborgne. Leborgne had already been in hospital at Bicêtre, Paris, for 21 years, but came to Broca's attention after developing gangrene. Leborgne had lost the power of coherent speech 21 years previously. He also suffered from epilepsy. After ten years in the hospital he had begun to develop paralysis on the right side and his vision had deteriorated. For the last seven years he had been unable (or unwilling) to get out of bed. Broca was a specialist in speech and Leborgne's disability interested him much more than did his gangrene. He wrote of the case: 'He could no longer produce but a single syllable, which he usually repeated twice in succession; regardless of the question asked him, he always responded: tan, tan, combined with varied expressive gestures. This is why, throughout the hospital, he is known only by the name Tan.'

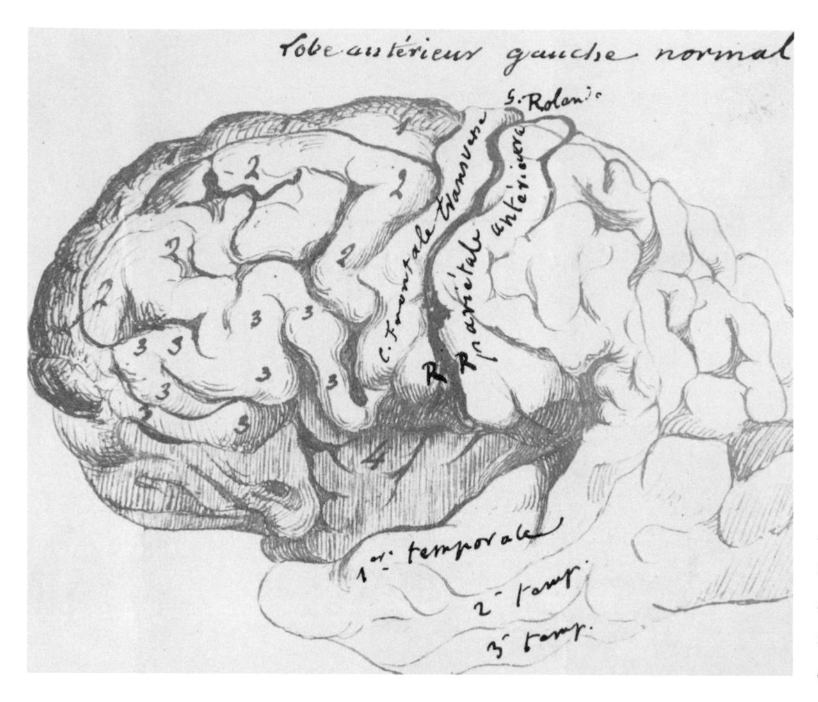

Broca's drawing of the brain, showing his thoughts on the location of the capacity for speech.

The loss of fluent speech is now known as Broca's aphasia. Sufferers often retain the ability to understand language, however, at least to some degree.

Leborgne soon died and Broca performed an autopsy, finding a large lesion in the frontal area (more precisely, in the posterior inferior frontal gyrus). A few months later, Broca saw another patient who had lost the ability to speak. Lazare Lelong, 84 years old, could say only five words. On his death, Broca found damage to the same area of the brain as he had found in Leborgne. He quickly concluded that use of language is localized to a particular part of the brain and, further, can be broken down into the ability to produce spoken words, to formulate utterances and to comprehend language. One of these faculties could be damaged without impeding the others, so localization was really quite detailed and specific.

Broca did not publish his full findings until 1864, by which time he had examined a further 25 patients/brains and could be certain that he was right. He was a highly respected physician and anatomist, so his findings were widely accepted where previous indications that speech might be localized at the front of the brain had been largely ignored. The reluctance of the medical profession to engage with anything that smacked of phrenology was dispelled by Broca's previous reputation and the strength of his evidence and reasoning. In addition, he was careful to point out that he was not saying speech was localized in the same place as claimed by Gall. Under Broca's championship, localization became acceptable and mainstream. The area he

identified as crucial to speech is still known as Broca's area.

Two years after the death of Leborgne, Broca pointed out that lesions affecting speech usually occurred in the left side of the brain, but in 1865 he said so with more conviction and precision. He continued to work on the problem with more patients, finally identifying four different types of speech loss and linking Broca's aphasia with damage to the posterior inferior frontal gyrus.

Learning again

Many of the patients with aphasia that Broca studied lived to tell the tale – literally. He discovered that within a few weeks, with suitable encouragement and therapy, some of his patients with speech loss learned to speak again. They were apparently co-opting a different part of the brain to do the work of Broca's area. His assumption was that it was the corresponding area on the other side of the brain. Modern ideas of the plasticity of the brain began with this finding of Broca's and his mute patients.

> **LEFT, RIGHT, LEFT**
>
> Broca encountered some patients who had a lesion in the right frontal lobe and attendant speech loss. He proposed two possible explanations: that if a person had already suffered damage to the left frontal lobe, the power of speech might already have relocated to the right side of the brain; or that in left-handed people the centre of speech is naturally on the right side of the brain. It is now known that Broca's area is on the left in almost all cases, regardless of handedness.

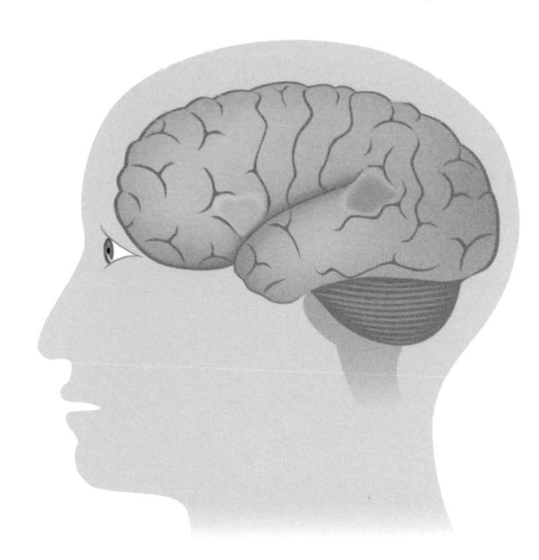

Broca's area is near the front of the brain (in blue) and Wernicke's area is nearer the back (in green).

Another area

Speech and its loss, it turned out, are more complex than they seemed. The type of aphasia described by Broca was certainly not the only type, and lesions of the frontal lobe were not always involved. In 1874, the German neurologist Carl Wernicke suggested that another area, near the back of the brain, was implicated in a condition now called Wernicke's aphasia. In this, the patient can still string together words in plausible sounding syntax, but understanding has been lost so the utterances are meaningless. This suggested that

two types of language ability and processing are involved in producing language – the physical articulation of sounds (Broca's area) and the connection of words with meaning (Wernicke's area). Wernicke drew maps of the brain connecting Broca's area with the area towards the back of the brain (in the superior temporal gyrus) that he associated with making meaning from and with language. This is now known as Wernicke's area, but recent imaging suggests it might not be quite (or only) where Wernicke located it.

Speech therapy can often help patients recover or improve their speech after traumatic brain injury.

BRAINS THROUGH TIME

Broca's decision not to cut into brains was a choice that greatly benefited later researchers, but limited his own discoveries. Leborgne's and Lelong's brains were preserved and have since been examined by modern neurologists. In 2007, Lelong's brain was scanned for the first time and Leborgne's for the third. From high resolution MRI scans (see page 177), damage was found to be more widespread than Broca had reported. In both brains, the lesions affected not just Broca's area but went as far as the superior longitudinal fasciculus, a thick bundle of nerve fibres connecting the frontal lobes to the back of the brain, including Wernicke's area. Damage to Broca's area was probably not the sole cause of the patients' speech loss.

THE PLASTICITY OF THE BRAIN

Called plasticity, the brain's ability to adapt and change lies behind recovery from brain damage. Neurons alongside those that are damaged have been shown to grow new connections and re-route the original processing paths. In some cases, the corresponding area of the opposite hemisphere can take on the functions of damaged parts, too. Recovery relies on rehabilitative therapy, which must begin early and be followed rigorously to reinforce the new pathways. Brain-imaging techniques can show which parts of the brain have taken over functions originally performed by the part that has been lost.

English neurologist John Hughlings Jackson (see page 39) was interested in the fact, noted even by Broca, that patients with aphasia are often able to curse fluently even when they lose the ability to make all other articulate utterances. He suspected that there is a distinction between automatic use of language, which might be centred in the right side of the brain, and thoughtful, intentional use of language, focused in the left (which could also manage automatic use). This meant that the two hemispheres were not wholly dissimilar, but the left took the lead in language. If lost, the right could pick up only the automatic use in which language is simply an emotional response rather than used to communicate meaning.

Jackson developed the idea that while volitional speech is localized at the front of the left side of the brain, perception and making meaning of perception (navigating, for example) are localized at the rear of the right side of the brain. He supported this with clinical evidence of patients who had suffered damage to the right side of the brain and could not recognize people or places. One of the patients conveniently died, and an autopsy revealed a lesion in the rear part of the right temporal lobe.

One side or the other

These findings fed into a major debate about whether or not the left and right hemispheres are functionally distinct. The earliest known speculation about this is found in an anonymous treatise thought to be based on the ideas of Diocles of Carystus, a Greek physician of the 4th century BC. He (or at least the anonymous treatise) claimed that 'there are two brains in the head'; that on the right is responsible for perception and that on the left for understanding. (The heart was also involved, in keeping with the prevailing beliefs of the Greeks at the time.) Even so, the thinking until the 19th century was that the two hemispheres were pretty much equivalent in form and function.

In 1865, Broca explicitly stated that speech is localized in the frontal lobe on the left-hand side. This had an impact on separate but related controversies: whether functions are localized in the brain, whether the two hemispheres are different or identical, and how they relate to each other.

Prior to Broca's findings, the assumption was that the two sides of the brain were equivalent and independent organs, just as the left and right eyes or ears are equivalent and can operate independently.

In two minds

The late 18th and early 19th centuries saw a new idea gaining ground. In 1780, Meinard Du Pui proposed that we have two minds, one in each hemisphere, just as we have other paired organs and body parts. In 1826, Karl Burdach suggested that the two parts are joined by the corpus callosum, a broad bundle of nerve fibres that runs between the two hemispheres. In 1840 Henry Holland, who was private physician to Queen Victoria, wrote about the double nature of the brain. He stressed that the bundles of nerves (called commissures) joining the two hemispheres act to keep the two halves working cooperatively. In this model, imbalance between the hemispheres or failure to communicate between them could lead to mental illness or madness. This interpretation relied on a physicalist view of madness – a physical dysfunction of the brain rather than a disorder of the spirit.

Who's in charge?

Contemplation of the two sides of the brain soon turned to the small difference in size typically found between the hemispheres and handedness (whether someone prefers to use their left or right hand). In the mid-19th century, the French physiologists Pierre Gratiolet and François Leuret claimed that the left hemisphere developed ahead of the right and weighed a little more during development. The assumption was that this

JEKYLL AND HYDE

In Robert Louis Stevenson's novella *The Strange Case of Dr Jekyll and Mr Hyde* (1886), the author explores the idea of two distinct personalities housed in a single individual. While the respectable Henry Jekyll is a well-rounded and mostly good, moral and intelligent individual, his alter ego Mr Hyde is gross, immoral, violent and entirely self-serving. The parallel with ideas about the left and right sides of the brain is clear. Hyde acts as though mad, and although initially his expression is under Jekyll's control, he eventually surfaces spontaneously and becomes increasingly difficult – and finally impossible – to suppress.

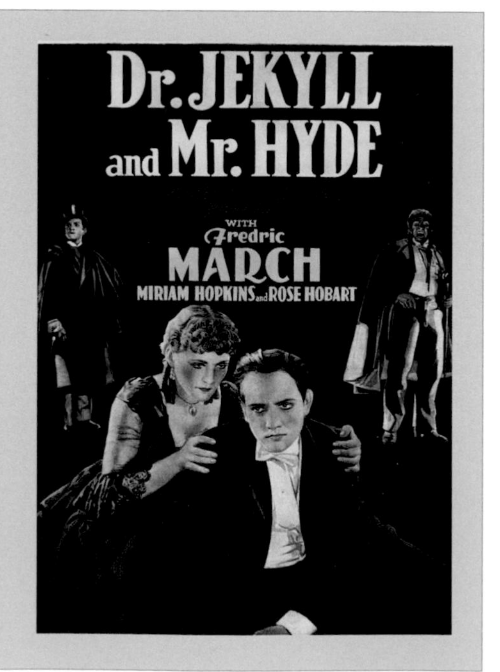

Being left-handed has attracted trouble in many times and places – unnecessarily.

gave it a head start in claiming dominance, so most people are right-handed.

During the last decades of the 19th century, brain asymmetry was even co-opted into shoring up the European white male's claims to superiority. It was commonly believed, and found by some studies of dubious rigour, that the degree of brain symmetry correlates with intelligence or mental development. So humans have asymmetric brains, but 'lesser' animals have increasingly symmetrical brains as we go 'down' the evolutionary tree. More provocatively, it was argued that 'inferior races' and women have more symmetrical brains than white men, and children have more symmetrical brains than adults. British physician John Ogle concluded that symmetrical convolutions of the infant brain must become asymmetric as the brain is developed through education. (It doesn't seem to have occurred to him that this undermines the point about the inferiority of women and other races, who had not benefited from the same education as white males.)

The interest in asymmetry led to many studies comparing measurements of both skull and brain to determine whether one hemisphere was usually larger than the other. Some found the left hemisphere to be marginally larger than the right.

All's wrong that's right

The consensus that a dominant and larger left side of the brain

LEFT-HANDED STUTTERERS?

After Philip Boswood Ballard stated that enforced change of handedness could lead to stuttering, numerous studies from the 1920s to the 1960s supported his findings. Then the theory was denounced as an urban myth (despite quite good evidence from the 20th-century studies) and fell out of favour. Recent brain imaging studies have resurrected the idea, finding that stuttering is related to a disturbed transmission of signals between the left and right hemispheres.

was the mark of the civilized and educated man put the right side of the brain in an embarrassing position. It suggested that a larger right side was the mark of an uneducated idiot or savage, which was exactly the conclusion that some people drew. In 1879, French neuroanatomist Jules Luys said that, in madness, the right side of the brain was larger than the left. He located base instincts in the right side of the brain. It became commonplace to associate the left side with morality and intellect and the right side with melancholy and irritation, at best, and with immorality, at worst.

Some people believed that the inequality of the hemispheres could be addressed, and that the right hemisphere could even be educated into better ways. Charles-Édouard Brown-Séquard, a neurologist from Mauritius, suggested a programme of education which encouraged children to use their left and right hands alternately and equally, as he believed an increase in use of the left side of the body would prompt the growth and (moral) development of the right side of the brain. Another more widespread practice was to try to force left-handed children to use their right hands instead. The first practice fell to criticism by James Crichton-Browne, who said that the history of civilization was founded on right-handedness and we should leave well alone. The second practice continued into the late 20th century, despite the publication in 1912 of evidence that forcing right-handedness was both unsuccessful and detrimental, and could lead to stammering. Interestingly, in 1906, J. Herbert Claiborne suggested that dyslexic children should be encouraged to develop left-handedness in an attempt to jumpstart the right side of the brain into taking over the lexical functions that the left-hand side was clearly messing up.

On and on

Once the principles of localization and distinction between the hemispheres were generally accepted, more and more abilities came to be associated with particular locations or sides of the brain. One important step on the way to further functional localization was the work of the German neurologist Korbinian Brodmann (1868–1918). Investigating the structure and histology of brain tissue, he drew up a map of 52 areas divided between 11 histological regions (that is, showing differences in the tissues). His method was based on the cytoarchitectural organization of neurons in the cortex – the arrangement of types of neurons, the way they lie and how they are stacked in layers. Subsequent neurologists have redefined and refined the

NOT JUST A MUSHY BLOB

With the development of better microscopy came new techniques using specialized dyes (stains) to show up the individual structures within cells. Anatomists began to discern differences between areas of the cortex. They mapped out 200 structurally discrete areas. The discovery that the brain is not structurally homogeneous reinforced the idea that functions are localized, with different structures supporting different types of brain activity.

areas, but Brodmann's remains the most widely used cytoarchitectural system. Subsequent studies have correlated several of the areas with particular functions.

As well as the principles themselves, the 19th century had established the practice of using dysfunction and autopsy as ways of understanding normal function. In the days before brain-imaging technologies, it was easiest to deduce which parts of the brain do what when they ceased or failed to do it. Correlation between cognitive or physical failures and brain lesions provided the only empirical evidence of brain function available. That would not change until the 20th century, when sophisticated brain-imaging technologies finally allowed detailed mapping of brain activity (see page 176).

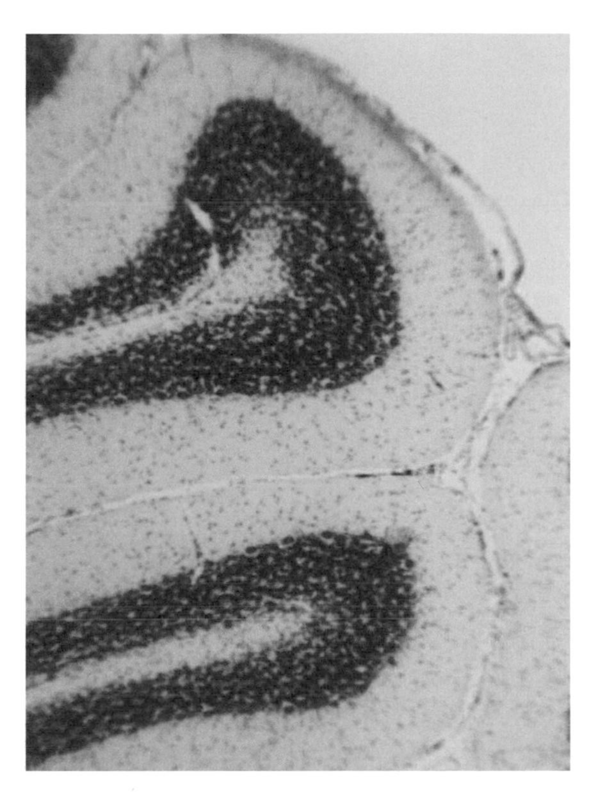

The use of Nissl stain on a sample of cerebellum from a cat clearly shows the cell bodies of neurons, which take up the stain.

THE LAZY BRAIN?

There is a commonly repeated myth that we use only 10 per cent of our brains. It's not clear where this originated, but it can be seen in advertisements and self-help brochures from the late 19th century onwards. It is frequently used now in similar contexts and to promote practices and products that claim to unlock or access extra brain power. But it is entirely a myth. Investigation of the brain with modern scanning techniques can find no significant part of it that is inactive under a range of stimuli, even in sleep. Damage to the brain causes problems at least in the short term, until routes around the damage are established (see page 62). No one would opt to lose 90 per cent of their brain on the basis that they would be just fine with 10 per cent. Finally, evolution would not allow us to waste so much of an organ that is so expensive to run: the brain is around 2 per cent of our body weight, but uses 20 per cent of the oxygen we consume. It is most certainly doing something with all that oxygen!

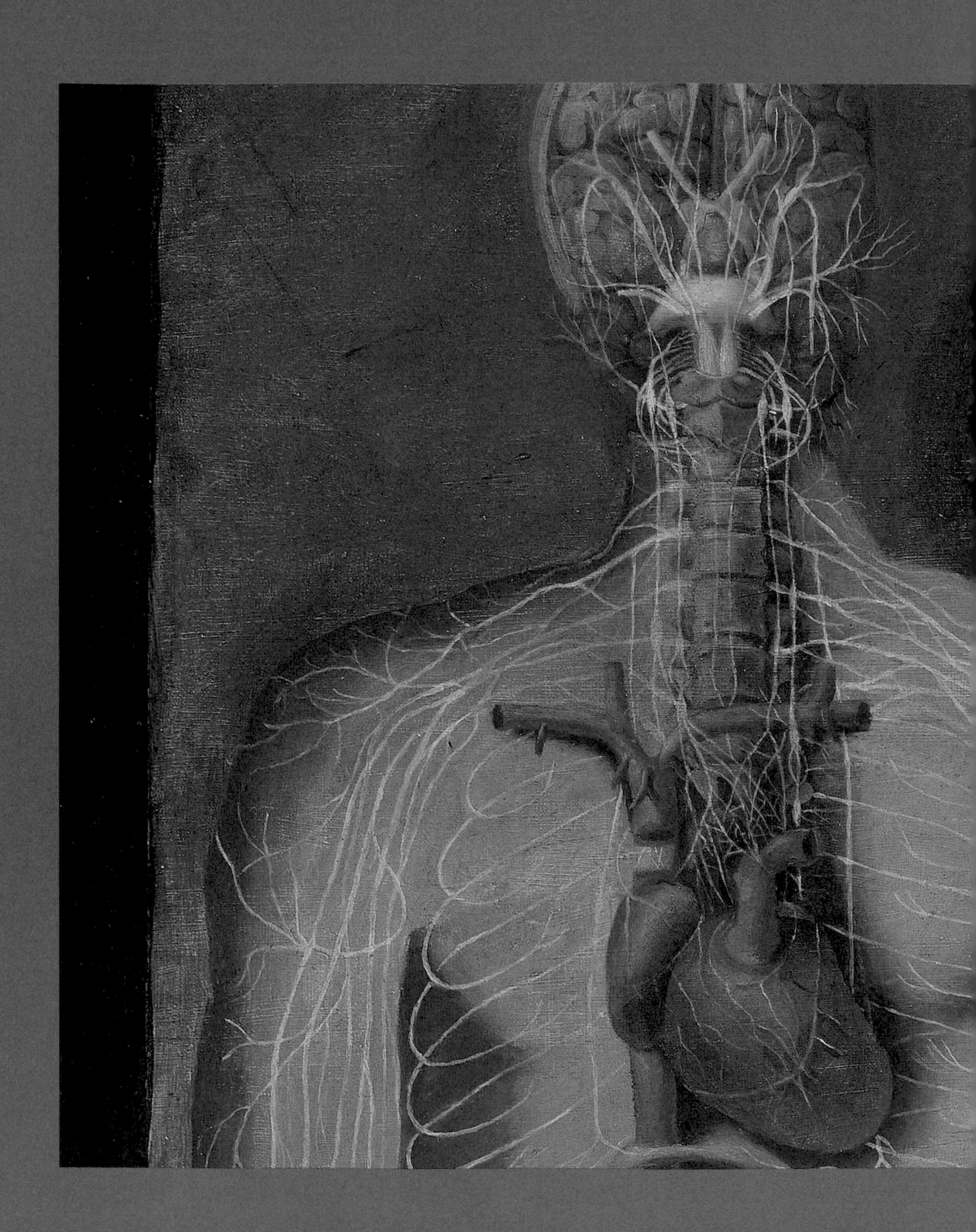

CHAPTER 4

A bundle of **NERVES?**

'If this Substance [the nerves] is everywhere Fibrous as it appears in many places to be, you must own that these Fibres are disposed in the most artful manner; since all the diversity of our Sensations and Motions depends upon them.'

Nicolaus Steno, 1668

It was apparent, even to Galen, that nerves were responsible for transmitting information to and from the brain. Exactly how they do it, though, was very difficult to discover.

A standing figure showing the vertebral column, nerves, heart and brain, painted by Jacques Gautier D'Agoty, 1765.

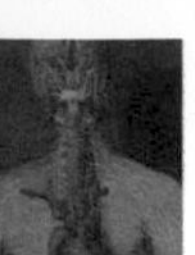

Communication network

If the brain controls the body and takes input from the senses, it must have some way of communicating with the rest of the body (and within itself). This mechanism is provided by the network of nerves which connects the brain and the spinal cord to all parts of the body. The nerves in the body – the peripheral nervous system (PNS) – are easier to study than the connections within the brain so were the first to be investigated. They can, with care, be dissected out to reveal large bundles of nerve fibres that can be seen with the naked eye.

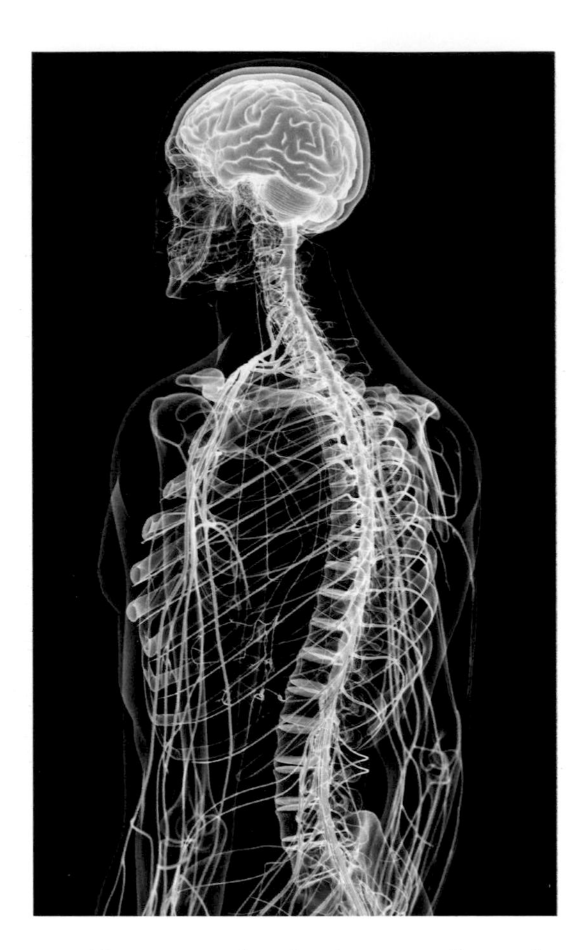

Nerves come out from the spine to serve all areas of the body; this network is called the peripheral nervous system (PNS).

Finding the nerves

Herophilus identified the peripheral nerves in the 3rd century BC, and Galen distinguished between two types of nerves, the sensory nerves and motor nerves, in the 2nd century AD. Galen considered the spinal cord to be an extension of the brain and noted that nerves branched from it and went to the limbs, where they could receive sensations or transmit the will to move from the brain to the muscles.

He believed – wrongly – that the nerves were hollow. This was not as a result of observation but because he believed the way they worked was to carry the animal spirits from the brain to the body, so they must be hollow. Galen saw the brain acting like a pump, contracting to push the *pneuma* through the ventricles from the front of the brain to the back and out into the motor nerves to the muscles. This explained the extreme speed at which nervous transmissions happen within the body, a speed he commented on – no time seems to pass between formulating the intention to move and actually moving.

Galen looked repeatedly for channels running through the motor nerves, but was unable to find any space within them. The only hollow nerve Galen ever found was the optic nerve of the ox. His notion of hollow nerves filled with *pneuma* moving to

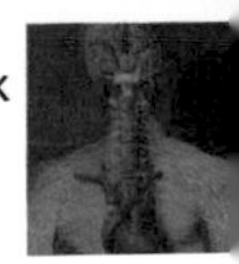

FROM INTENTION TO MOVEMENT

Several modern studies have found that there is no lag at all between intention and movement and that, in fact, the intention appears to come *after* the movement is initiated. MRI scans show repeatedly that the brain is already putting an action in motion before the subject is aware of having made a decision to act (see pages 197–9).

and from the brain would endure for 1,500 years, despite there being no evidence in the structure of the nerves to support it.

Impressionable minds and nerves of steel

Galen also considered the sensory and motor nerves to be physically different, their different functions clearly reflected in their structure. He thought the motor nerves were tougher as they had to carry the force of will from the brain to the muscles. They had to withstand pressure as the *pneuma* was squeezed into them by contraction of the brain. Those with the strongest will had the toughest nerves (hence the expression 'nerves of steel'). He believed the motor nerves originated towards the back of the brain and exited through the spinal cord.

By contrast, he thought the sensory nerves were soft, like wax, as they had to carry the impressions of the sense-object. These impressions were made on the nerves of the sensory organs, such as the eyes, and carried to the front part of the brain. The impressions from all five senses came together to be processed by the *sensus communis*, or 'common sense', which forged a perception of objects. Galen was confident that perception did not take place in the sense organs, as he knew from clinical experience that damage to the brain could sometimes impair sense perception, even when the sensory organs were healthy and undamaged.

Special nerves for special tasks

Galen's model persisted virtually unchanged into and through the European Middle Ages. Dissection and observation added some detail or refinement on the way, but made no significant changes to the central model. The Arab medical philosopher Ibn Sina, in the early 11th century, described nerves as 'white, soft, pliant, difficult to tear' and tried to trace their paths through the body and work out their various functions. But he, like Aristotle, saw the heart as the seat of control so he was already starting off on the wrong track.

Master Nicolaus, writing around 1150, left an anatomy text that usefully summarizes the beliefs current at the time about nerves and the brain. It is, inevitably, based very largely on Galen's teachings.

He repeats the familiar notion that the sensory nerves start in the *cellula phantastica* at the front of the brain, and the motor nerves in the *cellula memorialis* at the back of the brain. But he goes on to say that the sensory nerves are divided into five different types, which carry information relating to the five senses. Later, this was to become a point of considerable disagreement: are all the nerves essentially the same, but carry different types of information (or spirit)?

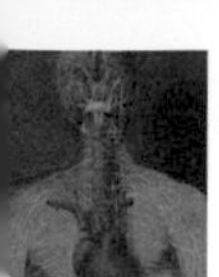

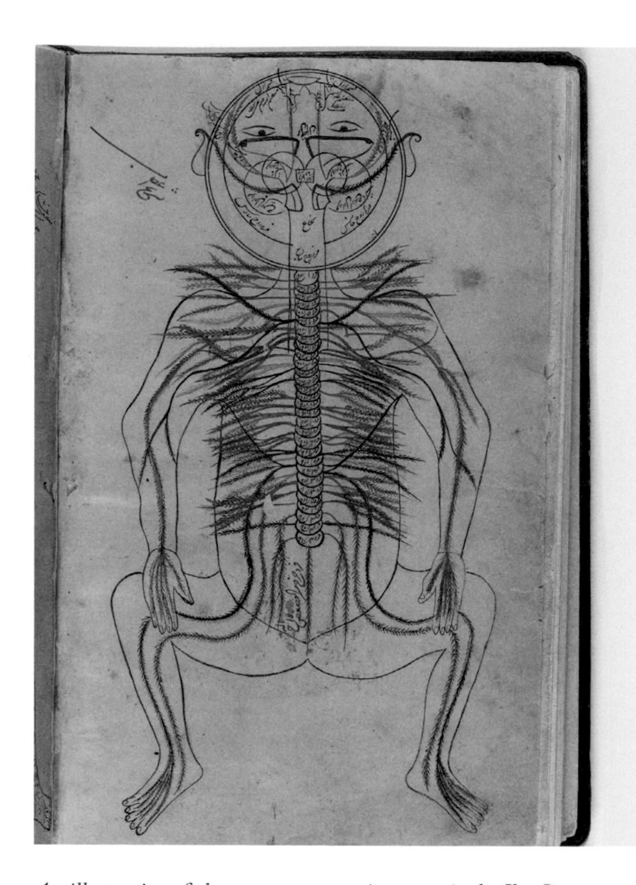

An illustration of the nervous system in a treatise by Ibn Sina.

Are the nerves different, specialized for the type of information they carry? Or is the 'information' the same, but sorted out in the brain according to its source?

Master Nicolaus was aware that nerves cross over, so those terminating in the left side of the brain serve the right side of the body and vice versa. He described two nerves emanating from the *cellula phantastica* going to the forehead, then crossing over, with the nerve from the left side of the brain connecting to the right eye and the nerve from the right side of the brain connecting to the left eye. There was a similar pair for the ears. (As he thought these nerves originated in the ventricle, he was not aware that the left hemisphere relates to the right side of the body, and vice versa – the nerves simply seemed to take a contrary path from the same source, the *cellula phantastica*.) He described the way nerves branch, so that a thick nerve goes to each of the arms and legs, but then splits into smaller branches to extend into the fingers and toes.

The motor nerves, according to Master Nicolaus, mostly serve to move the body, though he thought they had a minor function in detecting touch. Again, nerves cross so that they terminate on the opposite side of the body from their origin. He described nerves he claimed emerged in the cervical area

> *'By means of nerves, the pathways of the senses are distributed like the roots and fibers of a tree.'*
>
> Alessandro Benedetti, 1497

and where they terminated, and nerves that emerged in the dorsal area and the parts of the body they served. In particular, he noted that although the nerves were not responsible for producing speech they were needed to make the sounds that form speech. He said that a pair of nerves arose

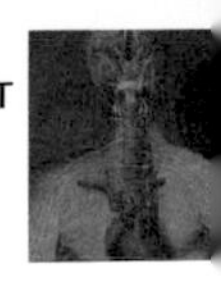

from the sixth dorsal vertebra, went to the lung and through the pulmonary region, then looped back to go to the tongue. If these nerves were too short, he reported, the person would not be able to produce the letter 'r'; if the nerves were too long the person had a lisp. This is possibly the first suggestion that a congenital condition of the nerves had a physical effect on the body's function.

The coming centuries would see the legacy of Galen challenged at last. Although for a long time the evidence of dissection was still interpreted in the light of Galen's account of the nerves, eventually that synthesis would prove unsustainable.

Conduits of thought

The means by which information travels to the brain or intention flows from it was never going to be easy to discover. Unlike the blood, which clearly moves through blood vessels and can be seen if spilled, the nerves have no apparent moving medium. The notion that the nerves carry animal spirit or a form of air persisted so long precisely because there was no good reason to reject it – cut open the body, and nothing can be seen moving through the nerves, but spirit could be invisible. That there was no visible channel was a hindrance, but not an insurmountable one, as it may just have been too fine to see.

Hollow or solid?

Galen assumed that the nerves were hollow as otherwise they could not have carried animal spirits. This is a fine example of a theory dictating a model even though it is not supported by evidence. In 1520, the

The accomplished 16th-century anatomist Andreas Vesalius first challenged Galen's account of the body.

Italian physician Alessandro Achillini wrote that 'the nerves are light to receive the spirit and thin in order to offer swift and easy passage to the spirit and flexible to serve the members.' He does not specifically say they are hollow, but nor does he deny it.

Finally the tide turned. Andreas Vesalius (see pages 26–7), a master of dissection and careful observation, was willing to contradict Galen, saying in his *De humani corporis fabrica* in 1543 that, 'I have never seen a channel, even in the optic nerve.' Furthermore, he pointed out that the *rete mirabile*, the network of blood vessels Galen described around the brain, does not exist in humans, and the ventricles are

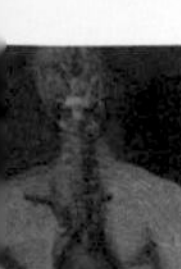

not as described by Galen. The following century, a medical student at the University of Edinburgh, John Moir, recorded in his lecture notes that 'nerves have no perceptible cavity internally, as the veins and arteries have.'

Yet despite this revelation, the idea of hollow nerves lived on in both the popular imagination and among many scientists. Perhaps one reason was that even though no channel inside the nerves was visible, there was no alternative suggestion – if the nerves did not carry *pneuma* or spirit, something like air, what did they carry? How did the nerves make the muscles move or carry an impression from the senses to the brain? It's difficult to reject a theory when there is nothing waiting to replace it; to move from a position of supposed knowledge to one of professed ignorance.

From automata to balloons

As we have seen, Descartes thought of the human body as a mechanism. As he still believed the Galenic model of hollow nerves, he could readily apply the principles of hydraulics and pressure which he observed in the automata of Versailles to the way nerves might work:

'Now as these spirits thus enter into the cavity of the brain, so they pass from there into the pores of its substance, and from these pores into the nerves; where, as they enter . . . now into some, now into other pores, they have the power to change the shape of the muscles in which the nerves are inserted, and by this means to cause motion in all the parts. Just as you may have seen that the power of moving water . . . is alone sufficient to move the different machines in the grottos and fountains of our kings' gardens, according to the various arrangements of the pipes conducting it.'

Descartes replaced the animal spirits model of Galen with a straightforward fluid which acted just like any other. Galen had never been explicit about the nature of the spirits – they were more an idea than an actuality, often thought to be weightless, for instance. Descartes, on the other hand, envisaged a very real physical substance with mass and volume. It might be a liquid, or perhaps a 'wind' or 'fine flame', but it followed the laws of hydraulics. Descartes thought the motor nerves carried fluid to the muscles which consequently increased in volume, producing movement. It's easy to understand how he arrived at this conclusion: just by bending your arm and watching the biceps bulge you can see the apparent increase in volume of a contracting muscle.

Descartes also gave the first account of how a reflex action might work, explaining it in terms of a stimulus causing the pulling of a tiny string that then opens a gateway in the brain, allowing fluid to flow into the

> *'[Statements by] this kind of people, who have never really examined the structure of the body, which is the work of God the Creator of all things, and who arrogate to themselves loose opinions taken from every quarter, are mere figments of the imagination, not unmixed with grave empiety.'*
>
> Andreas Vesalius, 1543

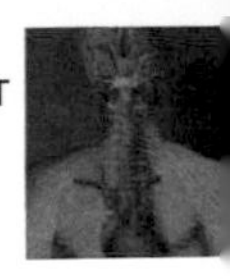

'If the fire A is close to the foot B, the small parts of this fire, which, as you know, move very quickly, have the force to move the part of the skin of the foot that they touch, and by this means pull the small thread C, which you can see is attached, simultaneously opening the entrance of the pore d, e, where this small thread ends . . . the entrance of the pore or small passage d, e, being thus opened, the animal spirits in the concavity F enter the thread and are carried by it to the muscles that are used to withdraw the foot from the fire.'

René Descartes

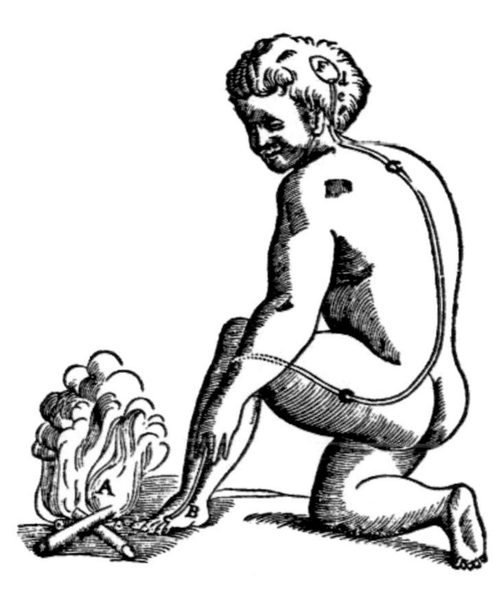

nerves and so to the muscles, causing the body to move away from the stimulus. In this way, the entire procedure is handled by mechanics, bypassing any need for the mind or soul (his *res cogitans*) to be involved. This is entirely appropriate, and even anticipates the account of the reflex arc handled in the spinal cord without recourse to the brain (see page 89). Although Descartes' explanation of how the reflex works was wrong, this first attempt at giving a physical cause for an automatic, unwilled action was a major step forwards for neuroscience.

Thomas Willis was a far more accomplished and important neuroscientist than Descartes, who was, after all, a philosopher rather than an anatomist or physiologist. The term *neurologie* first appeared in the English translation of Willis's book *Cerebri Anatome* (*Anatomy of the Brain*) in 1664. Even though Willis made many discoveries that could not be matched for centuries, and admitted he could not find any evidence of channels within the nerves, he still believed that they must be like 'Indian canes' (bamboo). His notion of how the nerves produced movement was slightly more complex than Descartes'. He proposed that animal spirit flowing into the muscle reacted with vital spirit to produce air. This inflated the muscle, making it swell, and caused movement. This became known as balloonist theory.

No holes and no spirit

Just a few years after Descartes' work, both the animal spirits and the balloonist theories were overturned by a simple experiment.

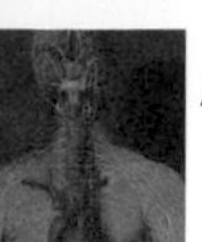

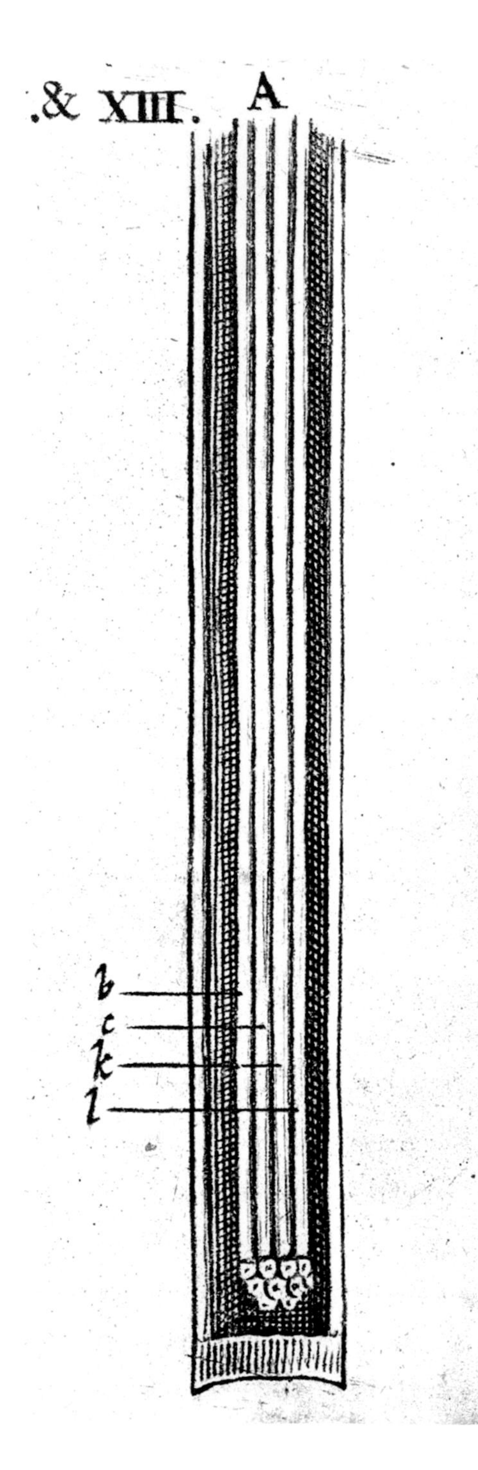

> *'The Nerves are nothing else but productions of the marrowy and slimy substance of the Brain, through which the Animal spirits do rather beam than are transported. And this substance is indeed more fit for irradiation than a conspicuous or open cavity, which would have made our motions and sensations more sudden, commotive, violent and disturbed, whereas now the members receiving a gentle and successive illumination are better commanded by our will and moderated by our reason.'*
>
> Helkiah Crooke,
> *Microcosmographia*, 1631

In 1662, Jan Swammerdam (see page 27) was dissecting a dog when he noticed that touching his metal scalpel to a nerve caused a muscle to contract. The muscle was not connected to the brain so could not have received any animal spirit from it. He tested his discovery by an ingenious method, which conclusively demonstrated that Descartes was wrong.

Swammerdam decided to measure the increase in volume of a stimulated muscle. He did this by dissecting out the heart of a frog and putting it into a syringe. He made sure there was an air bubble in the water near the end of the syringe and observed as the heart muscle (briefly) continued to contract and dilate. The bubble moved, demonstrating that the volume of the muscle really did

Descartes' idea of a nerve.

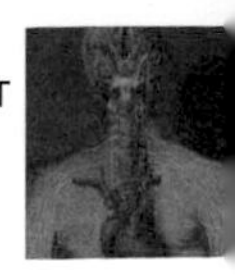

SPIRITS, FLUID, FIRE OR AIR?

Galen was not specific about the nature of the *pneuma* or spirit in the nerves, but it was certainly a physical substance, though refined. To early Christian writers, it was non-corporeal – more spirit in the ethereal sense. With the idea of a glandular brain, which emerged with Malpighi's observation of 'globules', it became quite specifically a liquid. Its distillation from the blood was thought to go through a process from ether, then nitro-ether, and finally to animal spirits (chemistry was still too undeveloped to devise an equation for the production of animal spirits). Surprisingly, considering the interest in the ventricles, which were supposed to contain this spirit, the cerebrospinal fluid that fills them was not collected and examined until the 18th century.

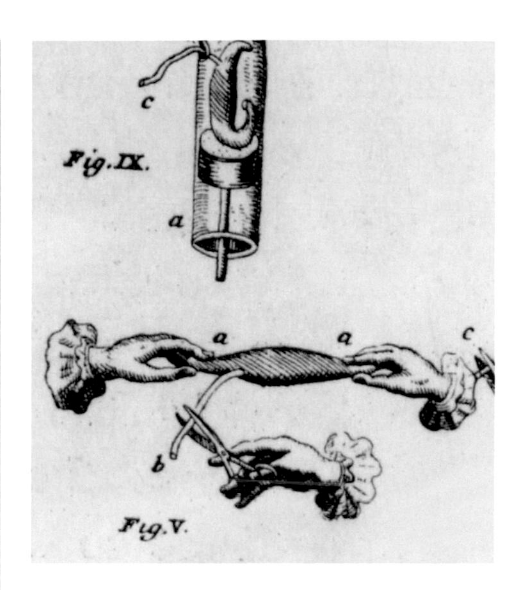

Swammerdam described an experiment in which he stimulated a frog muscle to make it contract: 'take hold, aa , of each tendon with your hand, and then, irritate b , the propending nerve, with scissors or any other instrument, the muscle will recover its former motion, which it had lost. You will see that it is immediately contracted and draws together, as it were, both the hands, which hold the tendons.'

change. But it contradicted Descartes' hypothesis – the bubble descended when the heart contracted, suggesting that the volume decreased where he had expected it to increase.

Swammerdam tried again using a different technique. This time he removed the frog's thigh muscle, along with its nerve, and put this into the syringe with the nerve protruding through a hole. He had already discovered that he could make a muscle contract by 'irritating' the nerve with a metal implement. This time, there was no discernible movement of the air bubble. This is what we would expect, as a muscle does not change its volume when it contracts. But it was not what Swammerdam expected or wanted. He explained away the result, saying that 'this experiment is very difficultly sensible, and requires so many conditions to be exactly performed, that it must be tedious to make it'. He argued that the muscle could not be expected to behave properly outside the body.

But ultimately this was not a result that could be denied. Swammerdam had demonstrated in his experiments that muscles contract following 'irritation' – a simple external stimulus – and not in

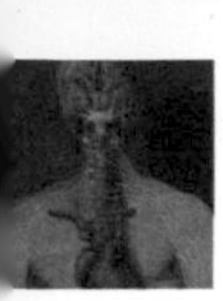

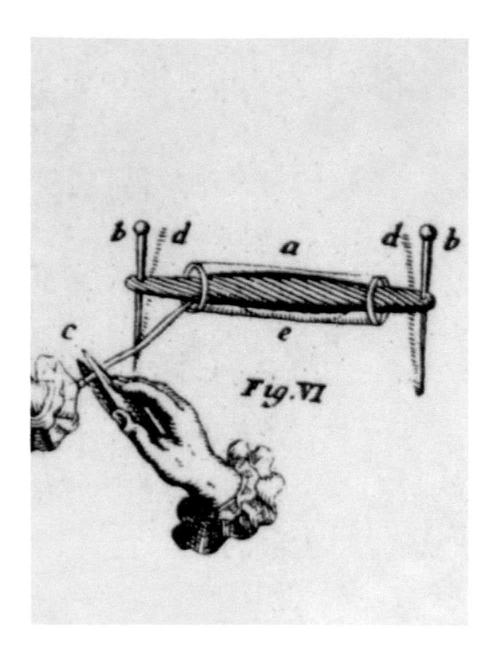

According to Swammerdam: 'If we have a mind to observe, very exactly, in what degree the muscle thickens in its contraction and how far its tendons approach towards each other, we must put the muscle into a glass tube, a , and run two fine needles, bb , through its tendons, where they had been before held by the fingers; and then fix the points of those needles.'

response to 'animal spirit' flowing through the nerves. His findings had wider implications, too. He had shown that just a part of an organism can be used to investigate mechanisms of the whole and that the mechanistic model of living organisms is valid. In showing that behaviour is a response to a stimulus it set the stage for later theories of learning, ranging from Pavlov's work on conditioning to the Behaviourist school of psychology.

Swammerdam's findings were confirmed by further experiments. The physiologist Francis Glisson (*c.*1599–1677) demonstrated that when a muscle is submerged under water and prompted to contract, the water level does not rise. From this, it's clear that the volume of the muscle doesn't change, so neither gas nor fluid is entering the muscle. Giovanni Borelli (1608–79), known as the father of biomechanics, carried out a very straightforward experiment which demonstrated that muscles are not inflated by gaseous animal spirits: he slit the muscle of a live animal while it was submerged in water. If gas had been pumped into the muscle, it would be expected to emerge as bubbles, but there were none.

THE MOVING BRAIN

If spirit, or nerve fluid, flows into the nerves, then something must make it move. Galen described the brain actively contracting to force the spirit into the hollow nerves. From the 16th century, anatomists debated this movement. Some claimed it was genuine and argued that the brain pulsates to drive the spirits, or even that the *dura mater* (the toughest of the meninges) constricts to squeeze spirits along on their progress. Others said that any observed movement of the brain was created by the flow of the blood through the arteries. Some anatomists claimed there was a relationship between the movements of the brain or spirits and the phases of the moon. Autonomous movement of the brain (or meninges) was only finally refuted by Thomas Reid in 1785.

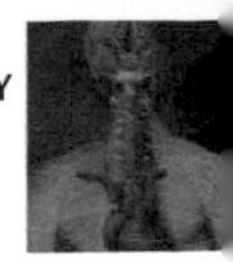

So, by the end of the 17th century, it had been fairly conclusively proven that nerves are not hollow, delivering animal spirits in the form of a fluid. Improvements in microscopy corroborated the absence of a channel running through the nerves. The Dutch microscopist Anton van Leeuwenhoek first prepared the optic nerve of an ox for examination in 1674. He reported: 'Schravesande . . . mentioned to me that since ancient times there has been some dissention among the learned about the optic nerve and that some anatomists affirmed [it] to be hollow. . . . I therefore concluded that such a cavity might be seen by me. . . . I solicitously viewed three optic nerves of cows, but could find no hollowness in them.' How, then, did they work?

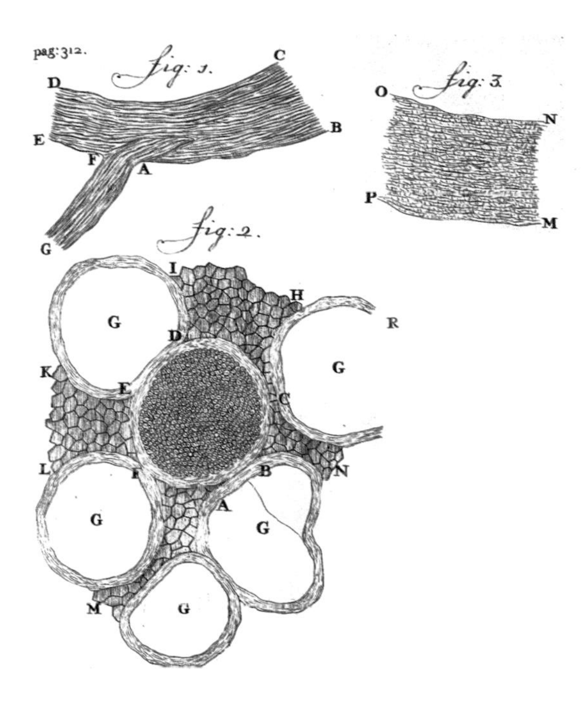

Anton van Leeuwenhoek's transverse section of a nerve, showing individual fibres, 1719.

Irritability and sensitivity

Before looking at other suggestions for how the nerves might carry signals, it is worth pausing to consider the word 'irritation' used by Swammerdam. The principles of irritation – here meaning simply stimulation – and sensitivity became important in classifying parts of the body and ultimately in distinguishing between the motor and sensory systems.

> *'From these experiments, therefore, it may, I think, be fairly concluded, that a simple and natural motion or irritation of the nerve alone is necessary to produce muscular motion, whether it has its origin in the brain, or in the marrow, or elsewhere.'*
>
> Jan Swammerdam, 1665

Francis Glisson, who had shown that the volume of a contracting muscle does not change, first developed the principles of irritability and sensitivity in the body. Not restricted to the nerves, irritability is found throughout the body, in the component fibres of all tissues and organs. Essentially, 'irritability' is simply susceptibility to being stimulated and responding to a stimulus.

Glisson divided the process of irritation into three stages: perception, when the fibre detects a stimulus; appetite, when the

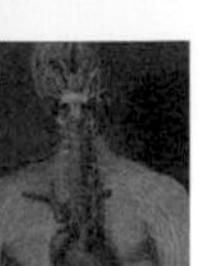

fibre is prompted or 'wants' to respond to the stimulus; and execution, when it carries out the required movement or response. Further, he divided irritation and response into three different categories according to the degree to which we are conscious of what is going on.

In many cases, the body just gets on with being irritated and responding. In digestion, for example, the irritated gut responds automatically and correctly by moving food through the intestines and digesting it. This Glisson termed 'natural perception'; the perception and response are handled locally, within the affected organ or tissue, he thought. In sensual perception, the brain is involved and there is communication between it and parts of the body by means of the nerves, but it is not a conscious response. The highest level, 'animal perception', is under conscious control and involves thought and volition. Glisson had essentially distinguished between the somatic nervous system, which deals with volitional (willed) movement, and the autonomic nervous system, which controls responses such as heart rate and breathing.

Glisson's work did not have much impact on scientific thinking because he framed it in terms of fibres perceiving a stimulus and developing an appetite for the required action. This endowed simple fibres or organs with faculties they do not have. The irritability/sensibility model became accepted only when stripped of these faculties. Consequently, the Swiss physiologist Albrecht von Haller (1708–77) is associated with the idea in place of Glisson. Haller was a brilliant physiologist and, while Glisson was given to theorizing and contemplation, Haller asserted nothing he could not back up with experimental evidence, relying on scalpel and microscope to reveal the workings of bodies. He narrowed Glisson's concepts, making irritability apply only to muscles and sensitivity only to nerves. Most importantly, he stripped out the need for body parts to perceive or develop an appetite, reducing the process to one of purely physical stimulus and automatic response. He defined irritability and sensibility as follows: 'I call that part of the human body irritable, which becomes shorter upon being touched; very irritable if it contracts upon a slight touch, and the contrary if by a violent touch it contracts but little. I call that a sensible part of the human body, which upon being touched transmits the impression of it to the soul; and in brutes, in whom the existence of a soul is not so clear, I call those parts sensible, the Irritation of which occasions evident signs of pain and disquiet in the animal. On the contrary, I call that insensible, which being burnt, tore, pricked, or cut till it is quite destroyed, occasions no sign of pain nor convulsion, nor any sort of change in the situation of the body.'

The sensitivity of nerves

Haller carried out experiments exposing different types of tissue and structure to stimuli including cutting, burning, noxious chemicals and blasts of air. His work was mostly carried out on dogs and cats. He found that in every case it was only nerves that were sensitive and those extremely so. He found the parts of the body supplied

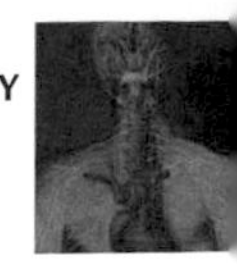

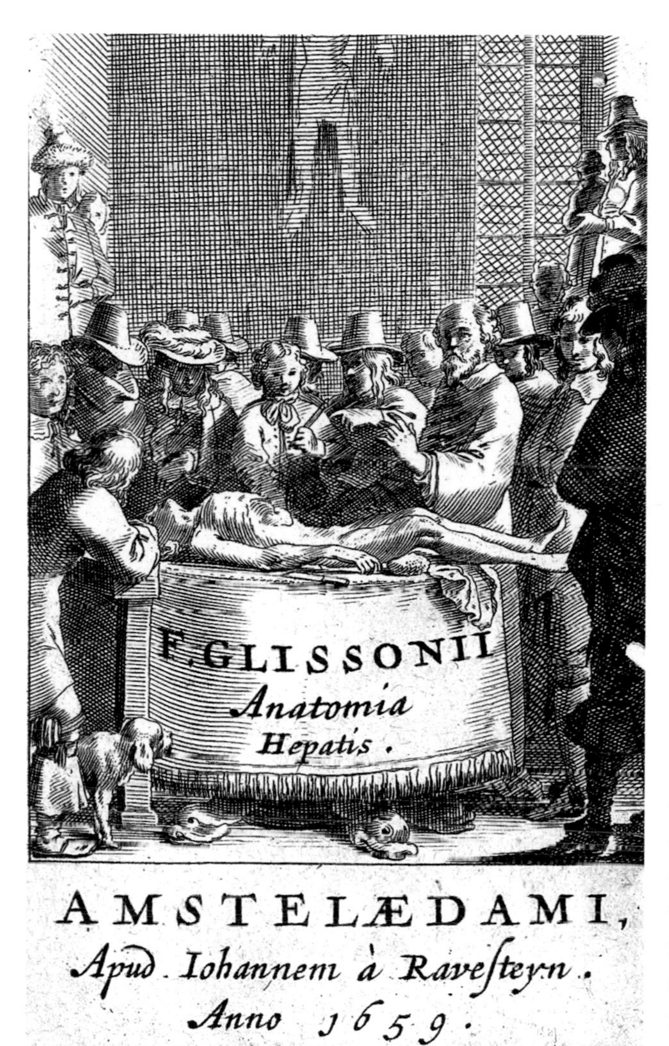

Glisson carrying out a dissection; the task probably afforded him opportunities to demonstrate the irritability of body tissues.

with most nerves to be most sensitive. And he discovered that if a nerve was cut, none of its branches below the cut responded, showing that the nerves do not interconnect to offer alternative routes of transmission.

The irritability of muscles

Haller found that stimulating a nerve made the muscles attached to it and any of its lower branches contract. The same happened even if the connection between the nerve and the brain had been severed or even if the animal was already dead. Most astonishingly, he discovered that 'irritability does not . . . arise from a nerve, but is innate in the fabric of the irritable part.' A muscle can be irritated and triggered to contract even when it is no longer connected to a nerve. He spelled out his conclusions: the least irritable parts are the most sensitive and vice versa; the nerves are needed to carry sensation to the 'soul'; irritating a nerve affects the muscles it is connected to, but causes no discernible change in the nerve; cutting a nerve removes all sensation below the cut, but does not impede irritability; 'irritability does not depend on the will or the soul'.

Haller's work was valuable in delineating the responses of muscles and nerves, clarifying that irritating a muscle either directly or through an attached nerve causes it to contract, and that a nerve (but not a muscle) is sensitive. His work did not, though, suggest exactly how the nerves

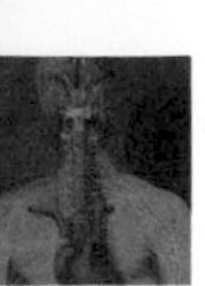

Albrecht von Haller carried out many experiments to discover which parts of the body would respond to stimuli.

might transmit information. He rejected the suggestion that vibrations might have anything to do with how nerves pass signals between body and brain (see below) and ended up with no alternative to the old notion of some kind of spirit or fluid travelling through the nerves.

'Buried in impenetrable darkness'

Swammerdam suspected that the workings of the nerves lay 'buried in impenetrable darkness' and it would be impossible to uncover them. Many people remained unwilling to abandon the animal spirits model completely, especially when there was no plausible alternative. The Dutch physician Hermann Boerhaave suggested that the nerve fluid is composed of very small particles, far smaller than those making up other body fluids. Consequently, it could travel through channels that are invisibly small.

Borelli, who had demonstrated that no gas comes from cut muscles, needed another way to explain the apparent expansion of a contracting muscle. He settled on a sort of chemical explosion occurring in the muscle, something like the bubbling 'effusion' produced when we mix vinegar and baking soda. The explosion, he proposed, was triggered by a drop of 'nervous fluid' or *succus nerveus* squeezed out of the nerves. The nerves were not hollow channels in his model, but were full of spongy pith that was turgid with nervous fluid. When the swollen nerve was struck or pinched, a drop of nervous fluid dropped out into the muscle at its end, initiating the 'explosion' that causes the muscle to contract.

Vibrations and vibratiuncles

Swammerdam himself wondered if the means of communication – which he recognized must be incredibly fast – could

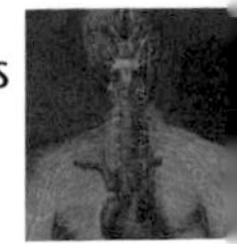

be similar to the way in which vibrations travel rapidly along a long beam that is struck at one end. The possibility of vibrations was suggested independently by some others. Borelli proposed that when the sensory nerves are compressed or struck, an 'undulation' travels along the nerve to the brain. Leeuwenhoek suggested something similar after seeing 'globules' within the optic nerve. He theorized that an impression made on the eye acts in the same way as a finger touching the surface of a glass of water and setting up movement of the liquid which is then transferred, globule by globule, from the retina to the brain. And the English philosopher David Hartley (1705–57) suggested that sensations are the result of the vibration of tiny particles in the nerves which are transmitted to the brain. Mild vibrations are experienced as pleasant but when vibrations are so strong that they break the continuity of the nerves, they produce pain. After a sensation, image or other sensory stimulus has passed, faint echoes of the original vibration persist in the brain, which he called 'vibratiuncles'. These provide the mechanism for memory.

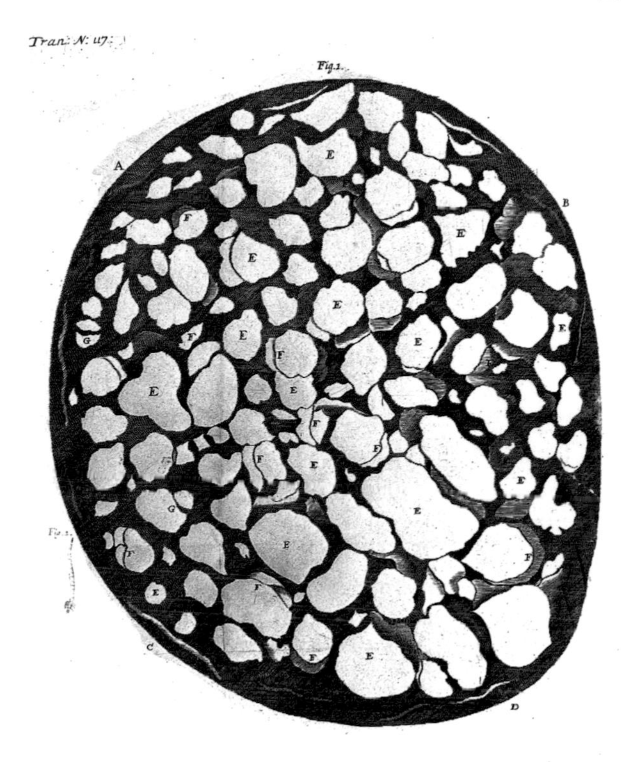

Leeuwenhoek's picture of the optic nerve showing the 'globules' that he thought passed an image to the brain.

However, as the anatomist Alexander Monro pointed out in 1781 and as Haller also indicated, the structure of nerves does not allow reverberations, vibrations or undulations to travel along them easily. When held taut, a string is good for vibrating, but the nerves, in Munro's words, are 'quite soft and pappy'. Some different type of mechanism was needed.

Frog soup and thunderstorms

While the notion of fluid 'animal spirits' flowing through hollow nerves had been thoroughly discredited, some physiologists, excited by the recent discovery of electricity, began to wonder whether 'electrical fluid' could flow through the nerves. Haller was one of those who thought the speed with which a stimulus is conveyed from nerve to

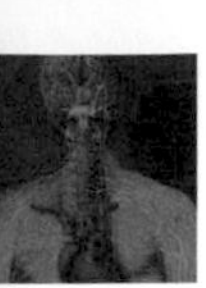

muscle suggested that electricity might be involved. The English physiologist Stephen Hales first suggested this in 1732, but no real progress was made with the idea until 1780 when Luigi Galvani (1737–98) had a little accident with a dead frog in Italy.

There are several versions of the story, but the most entertaining is that Galvani's wife was preparing frogs to make soup when she touched a metal knife to a leg and it twitched. Galvani experimented and found that touching a copper wire to the nerve and an iron wire to the muscle made the muscle contract. (A less domestic version has Galvani accidentally knocking his steel scalpel against the brass hook holding a frog's leg he was dissecting, making the leg twitch.) Galvani set about experimenting.

His most bizarre and conclusive experiment involved staking out frogs' legs during a thunderstorm and watching them jump. Taking the legs from a recently killed frog, he connected the nerves to a metal wire, the other end of which he fixed so that it pointed up into the sky during a thunderstorm. Every flash of lightning made the frog legs jump. This, Galvani claimed, demonstrated that muscles and nerves have an intrinsic electrical force, the action of which can be replicated after death by electricity from the atmosphere. It is this electrical force that causes nerve contraction and is the method of communication along nerves. He called his newly discovered force 'animal electricity'. He had conducted the first experiment in neuroscience and electrophysiology.

It was a bold claim. Not surprisingly, it met with some resistance. Galvani's countryman Alessandro Volta was among those who denied that there is any special 'animal' electricity and that the muscles were simply responding to the electricity produced by connecting the nerves between different types of metal. Galvani responded with more demonstrations, this time touching exposed nerves together or directly to muscles and producing the same muscle contraction with no metal or atmospheric electricity involved. He did this in 1797 by taking two frog legs with sciatic nerves extending out of the top and touching the nerves together, making the muscles of both legs contract. He believed that nerves had a nonconductive coating and the electrical impulses travelled along the centre of the nerves, finally entering muscles through tiny holes. This was astonishingly prescient, as we shall see.

> *'I am attacked by two very opposite sects – the scientists and the know-nothings. Both laugh at me – calling me 'the frogs' dancing-master'. Yet I know that I have discovered one of the greatest forces in nature.'*
>
> Luigi Galvani

Electricity triumphs

The proof that it is indeed a form of electricity which travels along the nerves and stimulates the muscles to contract came in the middle of the 19th century. German physiologist Emil du Bois-Reymond (1818–96) was a materialist, disinclined to accept any kind of aether or spirit in biology which could not be properly analyzed: 'no forces

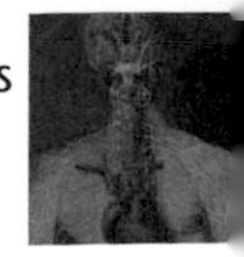

operate in the organism other than those common to physics and chemistry'. If an answer could not be found by looking at the known forces, he felt, then it was reasonable to assume an as-yet undiscovered force was at work, but only one 'of the same order as the physico-chemical inherent in matter'.

His interest in animal electricity began early, with a graduation thesis on 'Electric fishes', and shaped his work from the 1840s to the 1880s. He was inspired by the work of an Italian physiologist, Carlo Matteucci, who in 1830 began experimenting with a galvanometer and frog muscles. Matteucci showed that injured excitable tissue generates an electrical current, and that he could use this like a battery. He developed what he called a 'rheoscopic frog', or a frog galvanoscope, to detect electric current (see box on page 87).

Du Bois-Reymond devised his own equipment, including 'non-polarizable' electrodes, a generator and a potentiometer that allowed him to deliver graded bursts of electricity to his samples or subjects. A galvanometer measured and recorded the current produced through samples. Using this, he found and demonstrated what is now known as the 'action potential' of nerves in 1848.

His tendency to stage dramatic demonstrations probably helped his work to stick in the popular imagination. His most famous demonstration made use of his own body: he attached the leads of his galvanometer to his arms, then rested his hands in saline solution and waited for the needle of the galvanometer to come to rest. Then he would tense one of his arms, causing the galvanometer needle

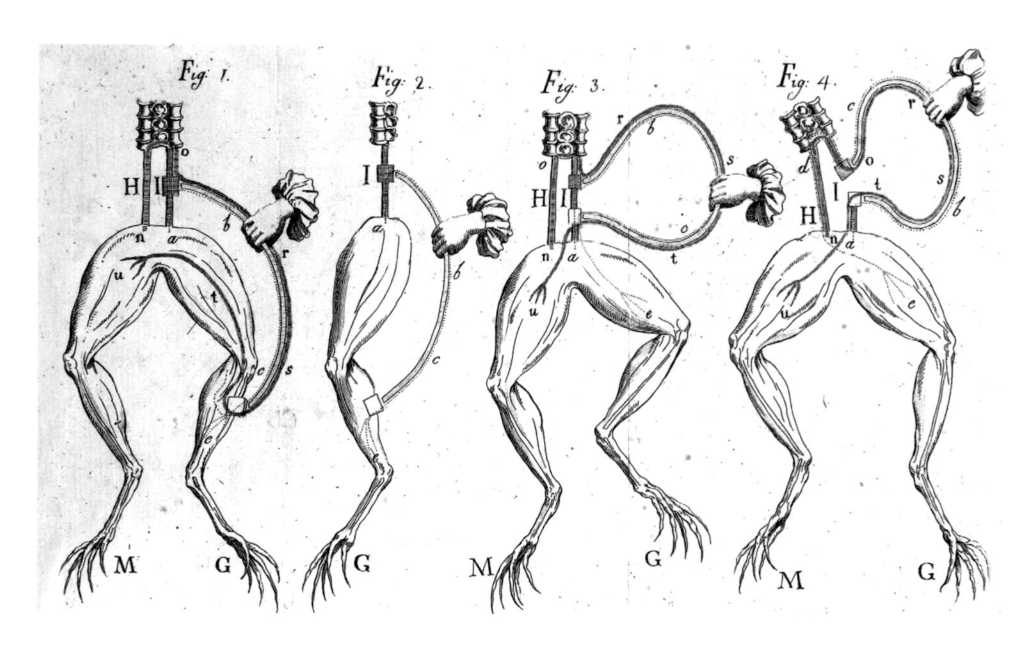

Galvani carried out various experiments and demonstrations using the sciatic nerves of frogs, stimulating them to cause the muscle to contract.

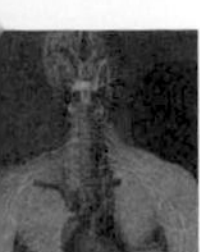

to jump wildly. It was a stunning and simple demonstration that he explained by reference to the electrical current produced in his own body as the nerves triggered the muscle action.

The speed of thought

While du Bois-Reymond's demonstrations were compelling, his model remained one among several in the contested territory of how nerves work. The question was only finally resolved in 1850 when the German physician and physicist Hermann von Helmholtz measured the speed of propagation of messages along a nerve. Using a frog (as usual), Helmoltz found that a nerve impulse took 1.5 milliseconds to travel 50–60mm (2–2.4in), a speed of 30mps (98ft per second). The modern value for transmission in frog nerves is 7–40mps (23–131ft per second). His measurement was inextricably linked to the method by which the impulse was transmitted, so finally resolved the question of the mechanism. The issue that remained for the 20th century was explaining exactly how the electricity is transmitted along a nerve.

FROM FROGS TO BATTERIES

Volta carried out his own experiments and discovered it was the presence of two dissimilar metals that produced the flow of electricity in Galvani's original discovery (the frog leg merely showed that the electricity was present). Volta went on to make the first electric battery, called the Voltaic cell; to do this he stacked brine-soaked paper between discs of two types of metal.

Feeling and doing revisited

We began this chapter with Galen's distinction between sensory and motor nerves, the one soft and impressionable, the other tough and hard. While there is nothing to support this view of the nerves, it survived for a long time, but finally disappeared, along with the explanation behind it – that the sensory nerves need to be soft to carry an impression and the motor nerves are turgid with animal spirit. With Galen's model rejected, the two types were left muddled together.

It was well known by the end of the 18th century that the nerves enter and leave the spinal cord to communicate with other parts of the body. It was generally believed, though, that the spinal nerves were of a mixed type and could simultaneously carry motor and sensory information and conduct impulses in both directions (to and from the brain). This potential tangle was sorted out by two men working independently at the same time, the neurologist Charles Bell in Scotland and the physiologist François Magendie in France.

In 1807, Bell wrote to his brother saying he had found that the sensory

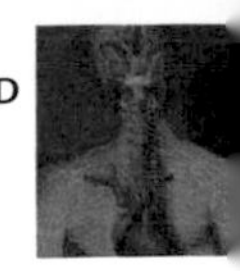

and motor nerves were of distinct types and went to different parts of the brain. In 1811, he wrote again, saying he had discovered that if he laid bare the roots of the spinal nerves, he could cut across the posterior nerve and it would have no effect on the muscles of the back, but if he cut across the anterior portion the muscles would go into spasm.

Bell made two serious career mistakes: instead of publishing his findings properly, he wrote them in letters to his brother and in a privately printed pamphlet, and he failed to state decisively that the posterior (dorsal) roots are sensory. This left the door open for a squabble about priority when Magendie also discovered the separation of motor and sensory nerves.

Cruel proof

Magendie carried out experiments on eight puppies he had been given. He was not, it turned out, the sort of person who should be given puppies. In 1821, he exposed the spinal nerves and cut one or more of the anterior and posterior nerve bundles, then applied the toxin *nux vomica* to try to induce convulsions. His experiment demonstrated the distinction between the nerve roots: 'The anterior and posterior roots of the nerves that arise from the spinal cord have different functions; that the posterior roots appear more specifically related to sensation, and the anterior to movement.'

Magendie's experiment caused extreme suffering to the puppies and was roundly condemned. He repeated this and other

THE FROG GALVANOSCOPE

The frog galvanoscope originates with Galvani's experiments and was improved by Matteucci. It comprises a skinned frog leg with electrodes attached to the nerve. If an electric current flows to the electrode, the leg twitches; it twitches again when the circuit is broken. The device was used to indicate the presence of a current (but could not measure the current). In fact, a frog galvanoscope is extremely sensitive and was used long after mechanical galvanoscopes and galvanometers became available. In 1848 the physician Golding Bird reported that the frog galvanoscope was 56,000 times more sensitive than a non-biological alternative.

To make one, the leg is removed from the frog, sometimes with a portion of sciatic nerve, and skinned. It is placed in a glass tube. Two electrodes are attached to the nerve, either one at each end or (more conveniently) both at the top but in different positions. The galvanoscope works best with a fresh leg; the leg must be replaced after 40 hours or so. (Don't try this at home – it is no longer legal and was never humane.)

Fig. 8.

The Galvanoscopic Frog.

BIO-BATTERIES

Another application of frog legs is the biological battery. Matteucci was the master of the bio-battery. His best battery was created from a series of half-thighs from frogs (the lower half of the thigh from each leg), but he also made batteries from ox heads, eels, half and whole frogs, rabbits and pigeons, and even one from living pigeons.

The thighs (right) are arranged end to end and placed on a wooden board (left) with dips filled with water and attached to electrodes at each end.

experiments in public lectures that produced no new findings. His disregard for his experimental animals prompted anti-vivisection legislation and his experiments were criticized by other scientists as 'abhorrent'.

Magendie was adamant that he should receive the credit for distinguishing the different functions of the anterior and posterior nerves, saying that Bell had come 'very near discovering the functions of the spinal roots'. In the end, the findings became known as the Bell–Magendie Law.

Further differences between motor and sensory nerves were not discernible until later in the 19th century when it became possible to examine not just the nerves, but the individual cells of which they were composed. It turns out Galen was correct – the motor and sensory pathways are structurally different – but not in the way he proposed.

The Scottish surgeon and anatomist Charles Bell opposed the cruel practices of physiologists such as Magendie in animal vivisection.

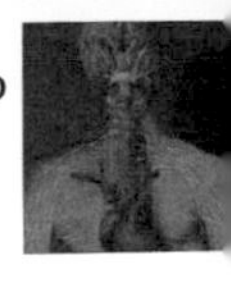

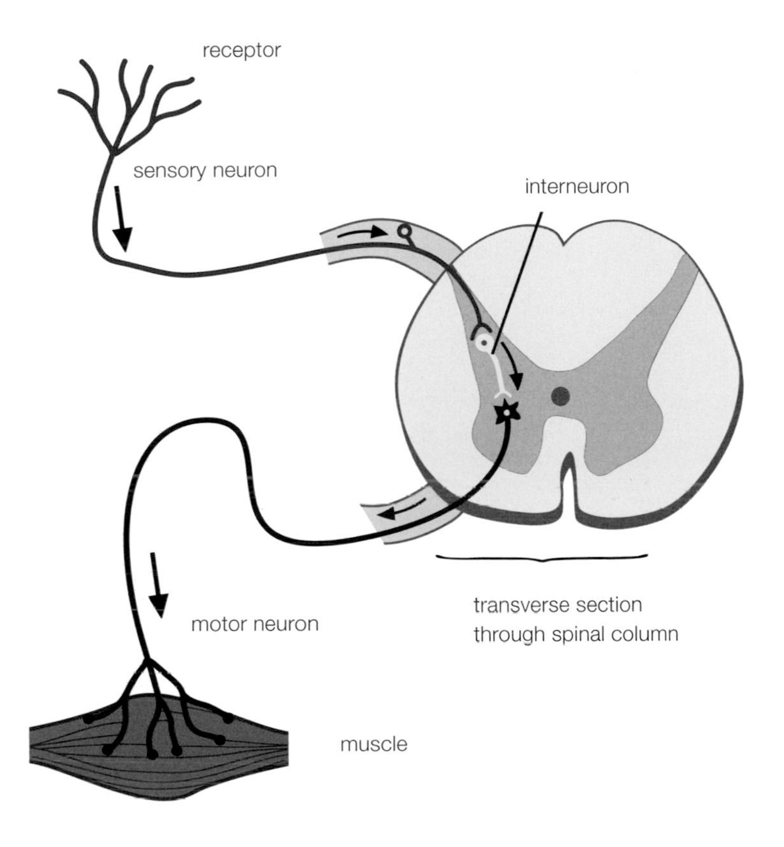

A reflex arc allows a stimulus to be processed and a response to be transmitted rapidly without involvement of the brain.

REFLEX ARC

Marshall Hall discovered that if he pricked the skin of a newt from which the head had been removed, the newt moved. This led him to his theory of the reflex arc, which had a stimulus and response being handled entirely in the peripheral nervous system, without recourse to the brain (which was absent, in the case of his newt). His argument ran: 'the spinal cord consists of a chain of units and that each of these units functions as an independent reflex arc; that the function of each arc arises from the activity of sensory and motor nerves and the segment of the spinal cord from which these nerves originate; and that the arcs are interconnected, interacting with one another and the brain to produce coordinated movement.'

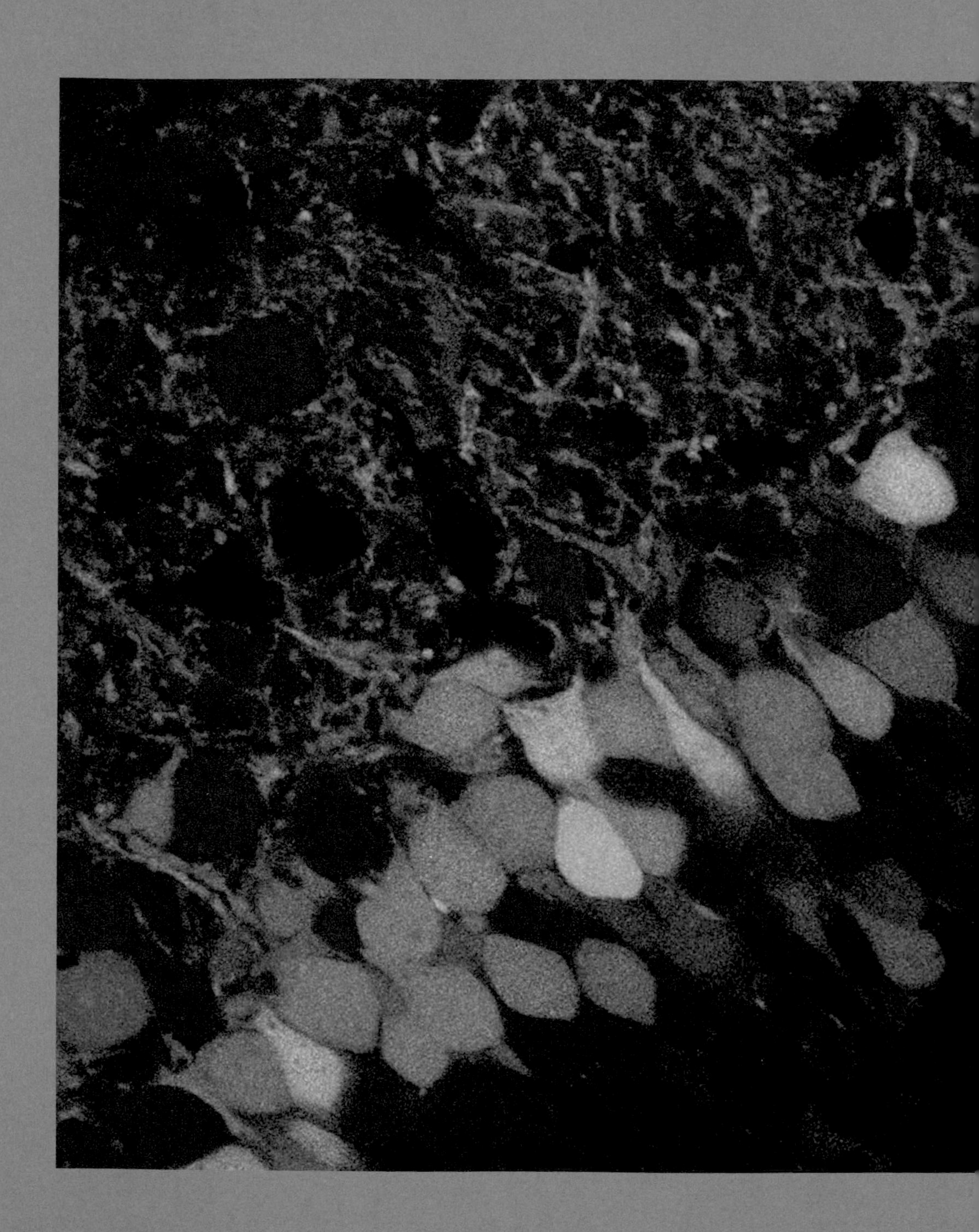

CHAPTER 5

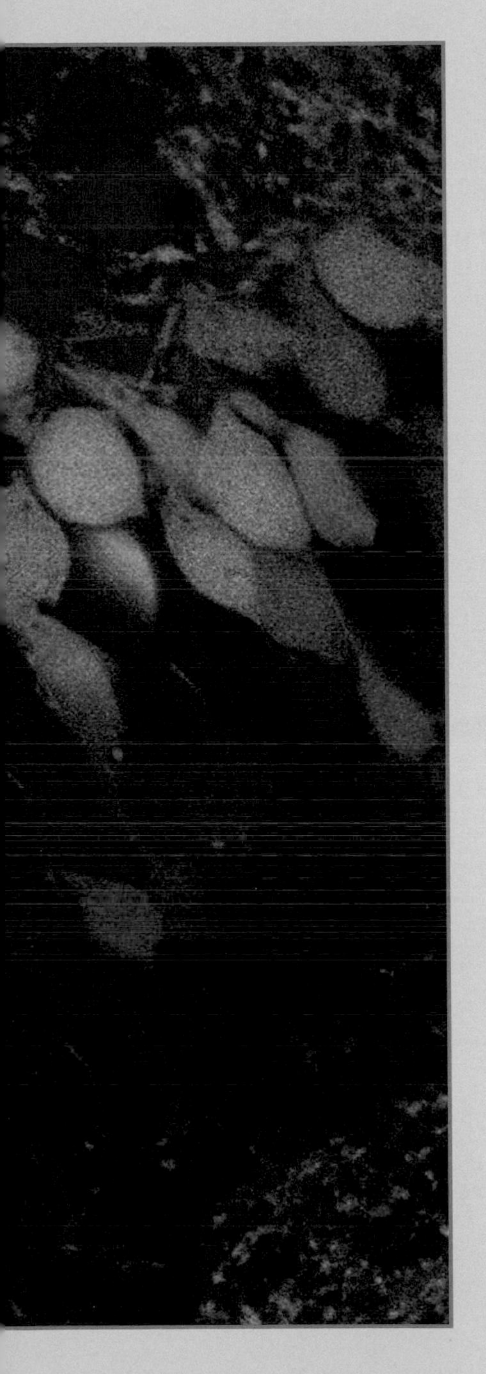

From fibres TO CELLS

'Often and not without pleasure, I have observed the structure of the nerves to be composed of very slender vessels of an indescribable fineness, running length-wise to form the nerve.'

Anton van Leeuwenhoek, 1719

From the time of Galen until the 18th century, nerves were observed in the peripheral nervous system and entering the spinal cord or brain. But with improvements in microscopy, nerves resolved into their constituent cells – neurons – and were finally observed within the brain.

'Brainbow' is produced by genetic engineering, introducing proteins that glow different colours and are used randomly in neurons. They help scientists to distinguish individual neurons. Brainbow is created in genetically modified mice using four colours which combine to make around 100 hues.

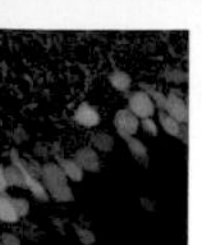

Seeing cells

As microscopy improved, the true nature of nerves slowly emerged, but the belief that they were hollow clung on long after it should have been abandoned, appearing in some textbooks as late as 1842.

Fibres and globules

Early in the 18th century, Leeuwenhoek described the nerves as filaments or threads held in bundles. This was soon confirmed by other microscopists. In 1732, Alexander Monro reported that 'nervous Fibrils' 'appear only like so many small and distinct Threads lying parallel, without any appearance of being Tubes'. He did, however, note that when cut transversely in their 'Interstices and Membranes' there are branches and openings which put observers 'in hazard' of believing they are seeing hollow vessels. In 1776, Italian natural philosopher Della Torre described peripheral nerves as rows of threads.

The son of Alexander Monro (also called Alexander) measured the diameter of nerve fibres in 1783, finding them to be 1/9,000th of an inch (about three millionths of a metre or 3 microns) across and that they appeared solid. More detailed examination of the nerves had to wait for improvements in the microscope itself, as well as in sample preparation and staining techniques.

Up to this point, the peripheral nervous system had been the major focus of study. The first person to record the examination of brain tissue was Marcello Malpighi, who reported seeing tiny glands or 'globules' with associated thin white fibres.

Examining tissue from the brains of various animals, Leeuwenhoek also reported seeing globules which were much smaller than those he had seen in blood (blood cells). As the brain was often considered to be a gland, or composed of glands, the globules were not considered surprising. It would be more than 100 years before the true form of nervous tissue in the brain could be discerned. Nervous tissue is difficult to work with because it

The human brain by Alexander Monro the younger, 1783.

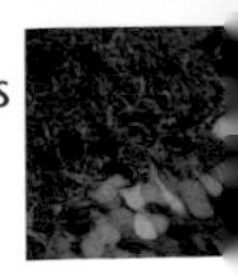

SLICING BRAINS

A sample examined through an optical microscope must be sufficiently thin for light to shine through it, usually no more than about 100 μm (a tenth of a millimetre). It's not easy to cut such slices by hand with a razor, particularly if the sample is as springy and stringy as a nerve. The development of the microtome, a device for holding a sample still and cutting thin slices from it, made the task much easier.

The first microtome was developed around 1770, and the technology soon improved. In early examples of the microtome, the sample to be examined was held in a cylinder, and slices were cut from the top by turning a crank handle. By around 1870, precision microtomes had appeared that consisted of a metal stage which held samples embedded in paraffin wax or celloidin (nitrocellulose) while a mechanically operated blade cut very thin slices. These serial sections made it possible to build up an impression of how structures were arranged and developed in three-dimensional space, which was vital for elucidating the structure of the brain.

A microtome is used to cut animal and plant material into slices for study under the microscope without damaging the delicate internal structure of the specimen.

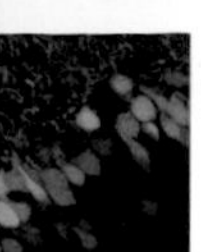

deteriorates rapidly, is not physically resilient and the parts are very small. Even when microscopy improved sufficiently to show the cells in other types of tissue, they were not immediately apparent in nervous tissue.

Cell theory

The notion that plant and animal bodies are composed of cells was put forward in the 1830s, but cells had been seen (and named) long before. Robert Hooke had drawn the cells that make up cork in 1653 and likened them to the rectangular cells inhabited by monks in a monastery. At this point, it was not clear that they were the basic structural component of all living things. This revelation emerged only in the 1830s.

In 1837, two German scientists, the botanist Matthias Schleiden and the zoologist Theodor Schwann, discovered they had each come to the same conclusion: that bodies in the domain they studied are made entirely of cells. Schwann published the finding in 1839, stating that cells are the basic units of life and that all organisms are made up of cells.

But nerves did not seem to contain anything that looked like the other cells of human or animal bodies. Some parts of nerve cells had been observed separately, but the connection between them was impossible to see using the microscopes and techniques available in the 1830s. It looked as though the central nervous system might be an exception to the rule which states that tissues are made of cells, and it was originally excluded from the general acceptance of cell doctrine.

Understanding neurons

Nerve cells – neurons – come in many shapes, sizes and configurations. This diversity delayed the recognition that all the various items being viewed under the microscope were comparable.

A 'typical' nerve cell has a cell body containing a nucleus, dendrites (filaments branching directly from the cell body) and an axon (a long filament attached to the cell body) with axon terminals (projections) branching from its end. But most nerve cells are far from typical.

The nucleus houses the important 'machinery' of the cell and is located in

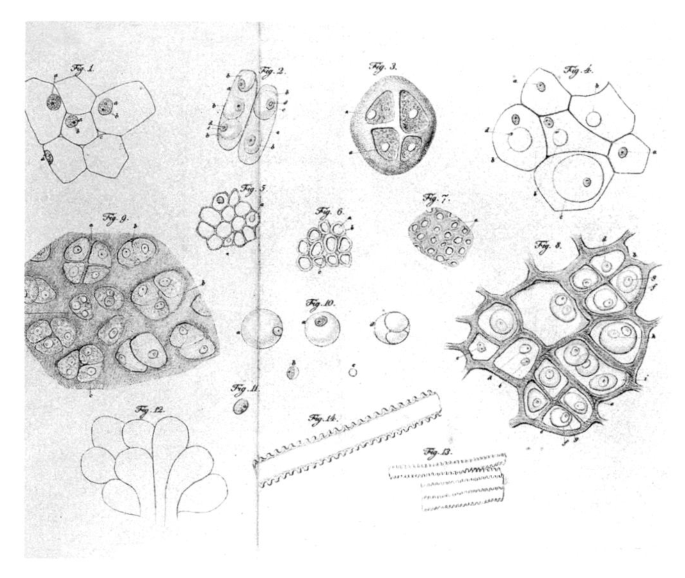

Schwann's drawing of cells, published in 1839 in his work establishing cell doctrine. Nerve cells were not included.

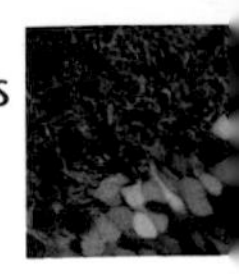

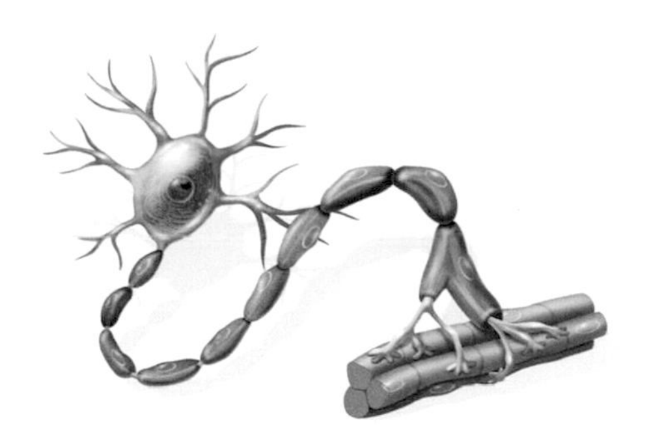

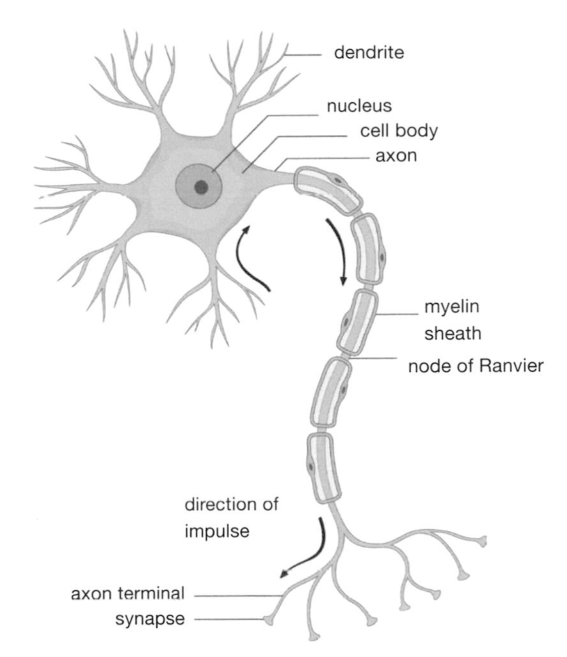

Above: a motor nerve cell attached to a muscle.

Right: a nerve cell showing its structure, and the myelin sheath.

the cell body. The axon terminal branches attach to a muscle (for a motor nerve) or end in a sense organ (for a sensory nerve). The axon can be short or long. The nerve impulse travels along it, either from the sensory organ to the dendrites, and then to another nerve cell, or from the dendrites along the axon to the muscle fibre.

The axon can have a myelin sheath, which acts rather like the plastic insulation on an electric cable: it insulates the axon, so the electricity of the nerve impulse is not dissipated. Myelin was first discovered in 1854 by Rudolf Virchow, a German physician. The myelin is not part of the neuron, but produced by separate cells, called Schwann cells, which wrap themselves around the axon. Gaps between the Schwann cells, called nodes of Ranvier, leave tiny points along the axon exposed.

The shapes and types of the neurons depend on their function and location within the body. There are three categories, but up to 10,000 different types within those categories. The categories are motor neurons, which relay information from the brain or spinal cord to a muscle to cause movement; sensory neurons, which carry information from a sensory organ to the brain; and interneurons, which carry information between neurons.

Looking at nerves

The German physiologist Gabriel Valentin was the first person to describe and draw part of a neuron in 1836. At last, the material of the brain was resolving into more than masses of white and grey matter – but it would be pieced together from the component parts of neurons rather than

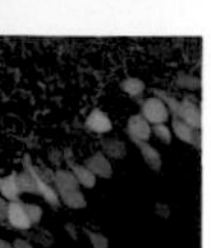

being revealed all together. Valentin's first drawings of neurons showed the nucleus and nucleolus, which together formed (again) a type of 'globule'. He also discerned very fine fibres which seemed to connect and wrap around the globules.

The same year, another German physiologist, Robert Remak, distinguished between myelinated and unmyelinated filaments, though at the time myelin was not recognized – it was just clear that some fibres had a kind of coat or wrapper and some did not. He also noted that nervous tissue is suffused by a network of very fine fibres or filaments.

Bohemian physiologist Jan Evangelista Purkyně (1787–1869) had wide-ranging interests and achievements. He was the first to notice that humans have individually distinct fingerprints and one of the first to make animated cartoons (an offshoot of his interest in the mechanics and neurology of vision and light). His most famous work relates to the nervous system. In 1837, Purkyně described clusters of drop-like cells, and a large number of fine, fibre-like processes found nearby – the same fibres that Remak had seen. Remak suggested the cells and fibres might be connected – that the fibres might emanate from the cells – but microscopy techniques were not good enough to confirm this.

The cells Purkyně drew, now known as Purkinje cells, are among the largest neurons and the easiest to see. They are found in the cerebellum. He described them in detail: 'Corpuscles surrounding the yellow substance [between the grey and white matter] in large numbers, are seen everywhere in rows in the laminae of

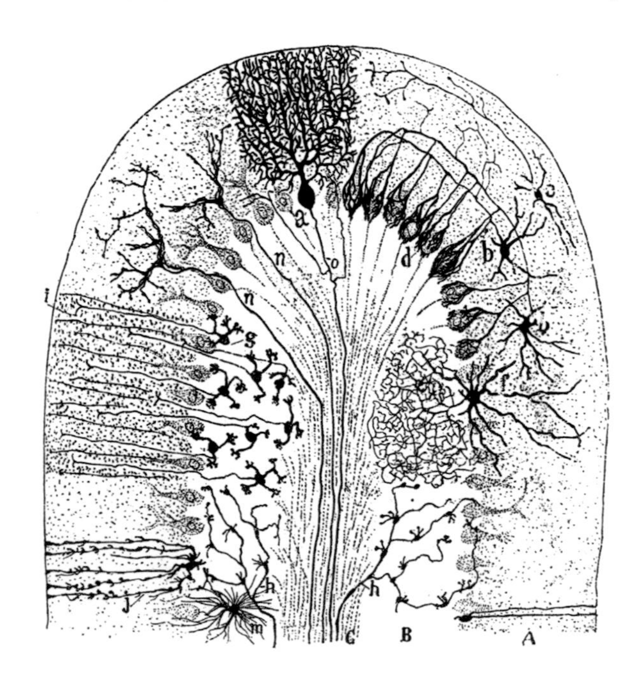

Spanish histologist Santiago Ramón y Cajal's drawing of different types of neuron in the mammalian cerebellum, 1894, shows some of the variety of shapes and forms of nerve cells.

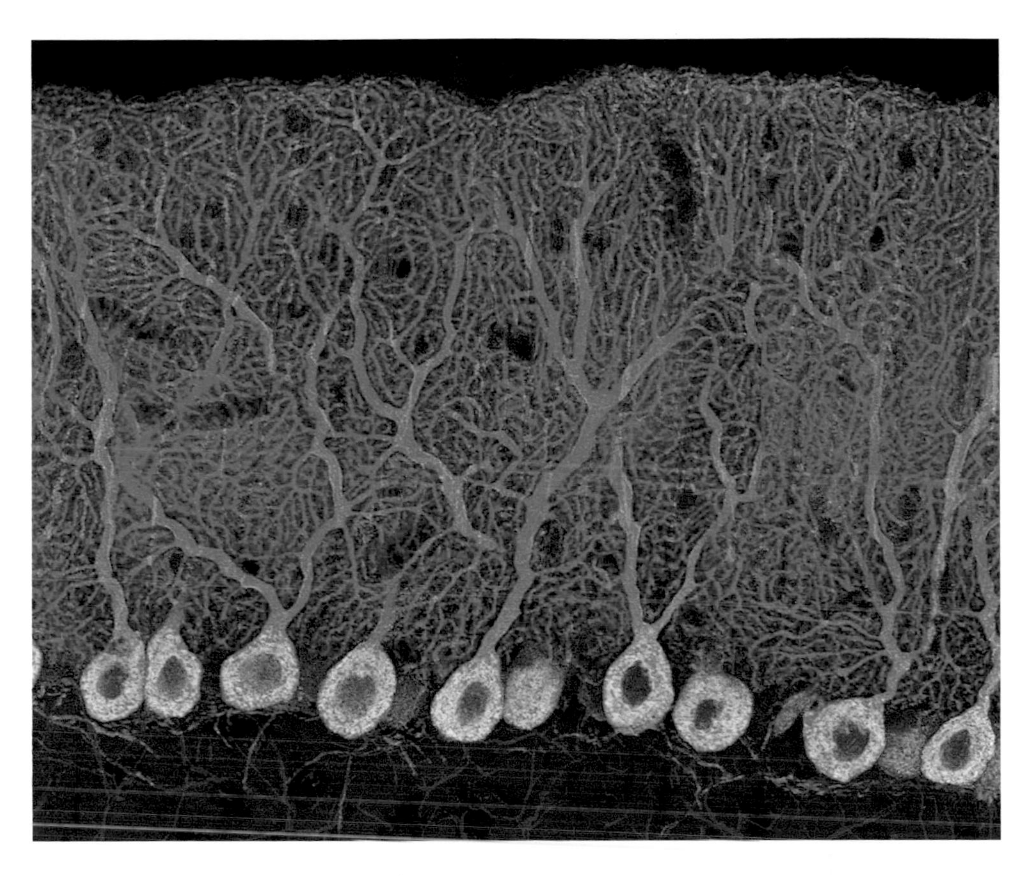

A false colour photo clearly shows the Purkinje cells in the cerebellum.

the cerebellum. Each of these corpuscles faces the inside, with the blunt, roundish endings towards the yellow substance, and it displays distinctly in its body the central nucleus together with its corona; the tail-like ending faces the outside, and, by means of two processes, mostly disappears into the gray matter which extends close to the outer surface which is surrounded by the pia mater [innermost membrane of the meninges].'

The part that 'mostly disappears' is the dendrites, too small for Purkyně to see and missing from his drawings, but clear in modern photographs taken with an electron microscope.

Long and stringy

It is difficult to distinguish between a mass of largely transparent greyish and whitish stuff, even through a microscope. This delayed the development of neurobiology in the 19th century until microscopes and supporting microscopy techniques improved.

The development of the microtome helped greatly, but it was still difficult to distinguish between the similar-coloured masses. Around 1863, Otto Deiters

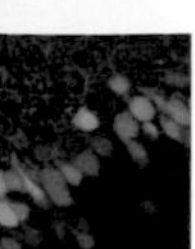

developed a microdissection technique which used chromic acid and carmine red as stains. The stains enabled him to isolate individual neurons for the first time. He identified different types of branching protuberances on each of the neurons he looked at. One type, consisting of short, tree-like branches, he called 'protoplasmic extensions' because they seemed to extend from the protoplasm of the cell body; they are now known as dendrites, the name given to them by Wilhelm His. The other was a long fibre with some very short branches at the end; he called this the axis cylinder, but it is now called the axon. He suggested that maybe the 'immeasurably thin' ends of the dendrites were fused to their neighbours, forming a large, unbroken network of filaments. Unfortunately for science, after such a promising start, Deiters died of typhoid fever at the age of 29.

The world in black and white

Deiters' stain helped, but it was not good enough for the level of detailed examination of neurons that would be needed to reveal the hidden world of the brain. Camillo Golgi, an Italian physiologist with a particular interest in the nervous system, made the necessary breakthrough in 1873.

Working by candlelight in a laboratory converted from a kitchen in the hospital for incurables near Milan, Golgi developed a staining method which he called the 'black reaction'. Now known as the Golgi method, it was achieved by adding a dilute solution of silver nitrate to samples hardened with potassium dichromate and ammonia. The silver nitrate stains the various parts of nerve tissue in different ways, making it possible to distinguish the different structures in the cells. It became immediately apparent to him that the cell body, axon and branching dendrites are all part of the same unit, confirming that nerves were, after all, cellular.

Network or neurons?

Discovering the different structures that made up neurons didn't help Golgi understand their function. He thought that dendrites provided nutrition. And,

PURKINJE CELLS AND AUTISM

The Purkinje cells are now associated with a number of functions including fine motor control, balance and proprioception (sensing the position of the body). Connections forged with Purkinje cells are responsible for developing motor skills such as playing a musical instrument or touch-typing. Nearly 80 per cent of children with autism spectrum disorder have problems with coordination and motor control. Research published in 2014 found a direct correlation between defective or abnormal connections between Purkinje cells and motor-skill deficits, suggesting that malfunctions of these cells are responsible for motor skill problems associated with autism.

observing a very dense and intricate network of branched axons in the grey matter of the brain, he concluded that the axons tangled into a continuous net. This was the theory he put forward when he published his first illustration of neurons in 1873. Since he considered the parts anastomosed (joined together) in a single large fabric, he didn't support the idea of localized brain functions. At most, he felt particular signals went to a broad area of the brain, but with it all interconnected true localization seemed unlikely. In 1871, German physiologist Joseph von Gerlach also proposed that the brain could be a 'protoplasmic network' – a vast, complex net, or 'reticula' – of nervous filaments. The reticulum of dendrites and axons soon became a common model of the brain.

But other physiologists, some working on animals, found evidence to contradict the fused reticular model. Wilhelm His, for example, studying the development of the central nervous system in embryos, concluded that the nerves are separate cells like any other, though varied in structure. The notion of distinct neurons could support the idea of localization of brain function. So the debate about whether neurons were separate or anastomosed became directly related to the debate about whether or not functions are localized in the brain.

Neuron doctrine

The real breakthrough in discovering how neurons operate together came with the work of the Spanish anatomist Santiago Ramón y Cajal, who first saw nervous tissue stained using Golgi's method in 1887. Cajal used an improved version of the technique, immersing the tissues twice in silver nitrate, to reveal structure which he recorded in astonishingly detailed and beautiful illustrations. (He had been a talented artist since his youth.) He discovered a truly astounding variety in the forms of neurons.

Ramón y Cajal's research developed into the 'neuron doctrine', published in 1894:

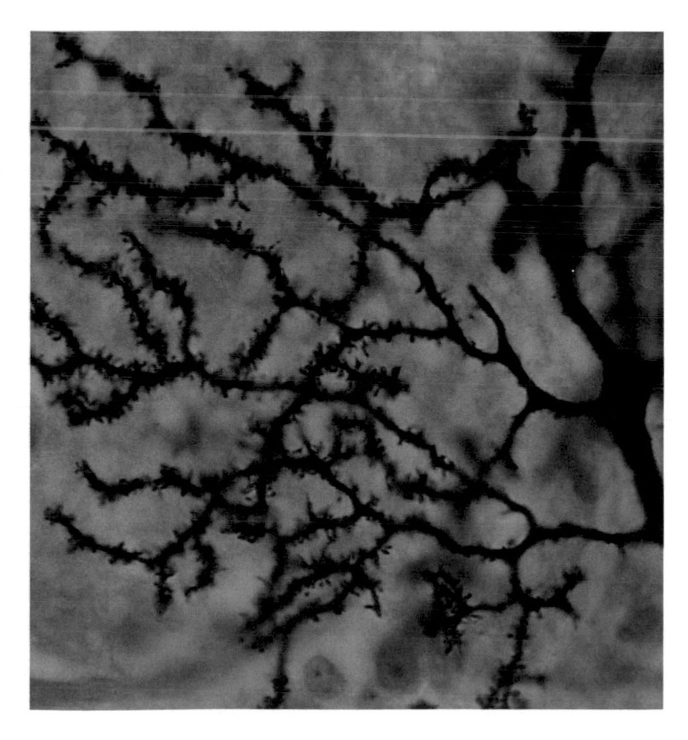

Dendrites stained using Golgi's method.

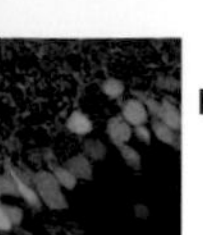

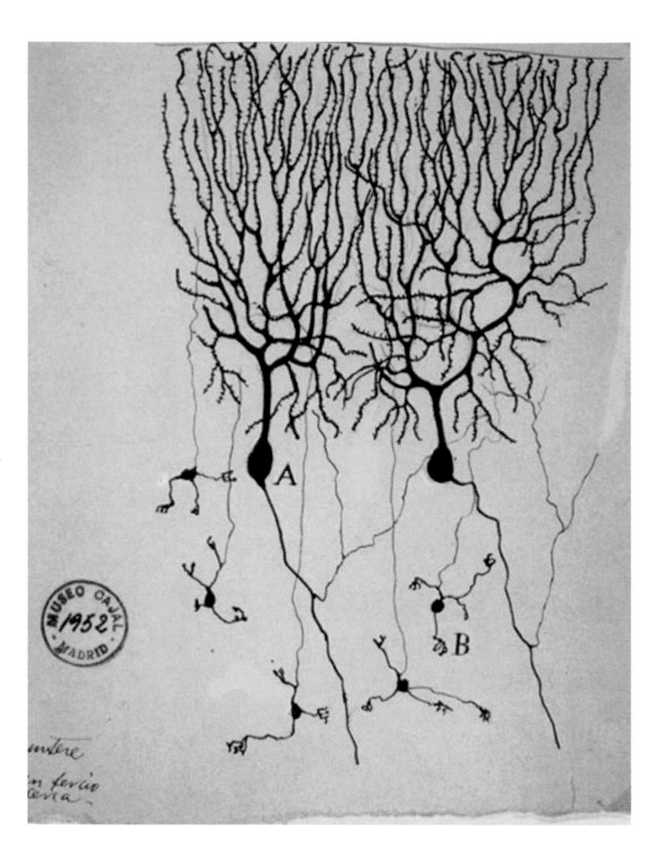

Ramón y Cajal's drawing of a Purkinje cell. While Purkyně had discerned only the blobby end of the Purkinje cell, Ramón y Cajal revealed the intricate complexity of the dendrites.

that the basic anatomical and physiological unit of the nervous system is the neuron, and that interconnected neurons provide the means of transmission of nervous signals. The term 'neuron' was coined in 1891 by German anatomist Wilhelm von Waldeyer, a great supporter of Ramón y Cajal, who even learned Spanish in order to read his original papers.

Bizarrely, Ramón y Cajal and Golgi shared the Nobel Prize for physiology in 1906 for their work on the nervous system, though they promoted directly conflicting views. Golgi did not consider the strands he saw to be separate nerve cells and argued vehemently against Ramón y Cajal's interpretation of them as such. Furthermore, Golgi was convinced that the function of the dendrites was purely nutritive. There was also personal animosity between them as Golgi had failed to gain much public attention with his original discovery of the new staining method and resented the glory heaped upon Ramón y Cajal. He felt it threatened recognition of his priority in uncovering the detailed structures, whether they were separate nerves or a network.

Mind in the gap

One reason that Golgi favoured a reticular model for the brain was that his stain was not effective with myelinated nerves. This made it difficult or impossible to trace individual

> *'I expressed the surprise which I experienced upon seeing with my own eyes the wonderful revelatory powers of the chrome-silver reaction and the absence of any excitement in the scientific world aroused by its discovery.'*
>
> Santiago Ramón y Cajal, 1917

NEURONS, NERVES AND BUNDLES

Neurons are single nerve cells. A nerve can consist of many neurons linked end to end. Typically, the dendrites at the end of one neuron are in close (but not fused) contact with the axons of another neuron, linking them in a chain which might go from a sensory organ to the brain, from the spinal cord to a muscle, or from one location within the brain or spinal cord to another. Nerve fibres are often bunched together in bundles. It was these bundles of fibres that were first observed, initially with the naked eye by Herophilus and those who came after him, and at a finer level by the early microscopists.

neurons through the network. Ramón y Cajal not only improved Golgi's technique, but investigated tissue from different types of animals, including the brains of birds, which have more non-myelinated cells. (Golgi restricted himself to human tissue.)

In examining the cerebellum, Ramón y Cajal saw tiny gaps between the tips of the axons of basket cells (a type of interneural neuron found in the cerebellum) and the cell bodies of adjacent Purkinje cells. It was clear to him that the cells could never be fused. He gave several good reasons for his own model, including evidence that if a neuron is cut, degeneration does not spread beyond the individual neuron to others, which it would if they were fused into a single unit.

So it emerged that the neurons are not fused together but approach each other very closely. The spaces between the neurons are as important as any neurological structure in how nerves transmit information. Yet exactly how these gaps – named synapses

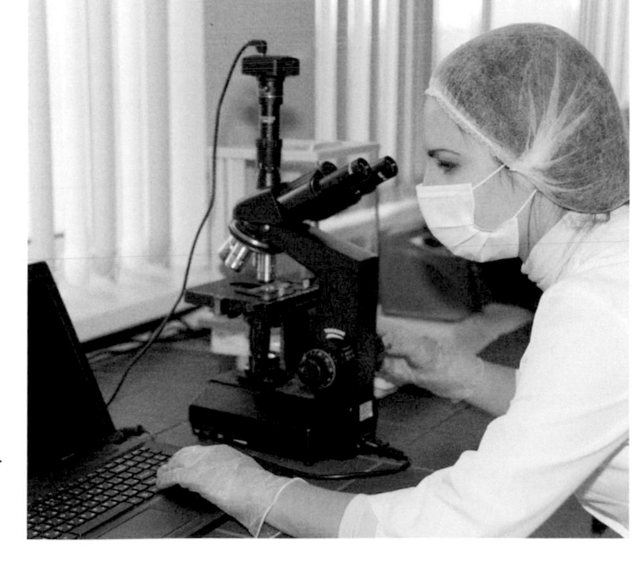

An electron microscope uses beams of electrons instead of light to resolve an image. As the wavelength of the beam can be much smaller than the wavelength of visible light, an electron microscope can produce greater magnification and detail than an optical microscope.

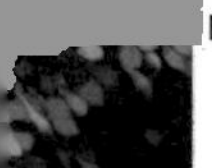

SANTIAGO RAMÓN Y CAJAL (1852–1934)

As a boy, Santiago Ramón y Cajal was deemed rebellious, disobedient and difficult. He was imprisoned at the age of 11 for destroying a neighbour's gate, which he did by firing a cannon he had fashioned himself. He was a skilled artist and gymnast, but his talents were not encouraged and instead he was apprenticed to a shoemaker and barber. Later, hoping to turn his son into a medic, his father took him to graveyards to search for human bones to examine. Sketching these bones was a turning point and inspired Ramón y Cajal to train in medicine. He served for a while as a medical officer in the Spanish army, but contracted malaria and tuberculosis while in Cuba in 1874–5. On recovering, he switched to teaching anatomy in Valencia.

He first studied diseases and epithelial cells (cells that form thin linings in the body), but turned his attention to the central nervous system after discovering Golgi's staining method in 1887. He made extensive studies of nervous tissue in many species and in all areas of the human body. Ramón y Cajal's artistic skill contributed to his success, aided by his ability to visualize in three dimensions and see how the slices he examined might fit together.

In stating the first tenets of the neuron doctrine, Ramón y Cajal set the scene for the development of modern neuroscience and has often been called the 'father of neuroscience'.

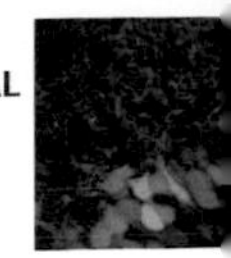

in 1897 – fitted into the picture was not clear until the 1950s, when they could be examined using an electron microscope.

Electricity is chemical

By the end of the 19th century, then, two principles were in place that needed somehow to be brought together. First, that the nerves carry some form of electrical impulses, as demonstrated by Emil du Bois-Reymond. Second, that the central nervous system relies on cells called neurons found in huge numbers in the brain and spinal cord and, less densely, connecting these to the body's sensory organs and muscles.

There was tension between those who favoured a mechanistic model of the nervous system and those who left space for something more ethereal – a spirit or soul. The latter were prepared to accept that electricity played a role as it was not, at the time, explicable in physical terms. The former looked to the tiny gaps between nerves with the intention of discovering exactly how an impulse could move from one to another. These researchers looked for a chemical means of transmission.

Poison arrows

The first clue came from a study by French physiologist Claude Bernard (1813–78). He wanted to understand the working of curare, a poison used by indigenous hunters in South America. Prey-animals (or enemies) struck by darts dipped in curare became paralyzed and then died of asphyxiation. In 1844 Bernard showed that the poison blocks the passage of a signal in the motor nerves, an effect he described as 'intoxicating' the nerve. His pupil, French histologist Alfred Vulpian, would go on to prove that curare acts at the point of connection between the nerve and the muscle. Vulpian proposed that there was a gap between the nerve ending and adjacent muscle cell even before Ramón y Cajal demonstrated one. Vulpian suggested a chemical means of transferring the impulse from nerve to muscle, and this would be the point at which curare would intervene.

Strychnos toxifera, the source of curare.

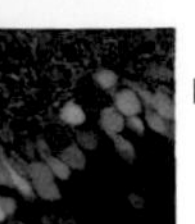

NEUROTOXINS

Curare is now classified as a neurotoxin. Neurotoxins work in many different ways to block or interfere with the transmission of signals along or between neurons. Curare is classed as a receptor inhibitor, which means it prevents the reception of a neurotransmitter chemical (nicotinic acetylcholine, in this case) at the junction between a motor nerve and a muscle. This prevents the flow of sodium ions into muscle cells which is needed to trigger muscle contraction. By preventing muscle contraction, curare stops muscular movement in the body, including breathing. Today, controlled doses of curare are used medicinally as a muscle relaxant.

A chemical 'pressor'

Curare was not the only chemical found to have an impact on nerves and muscle. In the late 19th and early 20th centuries, several scientists experimented with an extract from the adrenal gland, first studied by the Polish physiologist Napoleon Cybulski in 1895. It contained the chemical now known as adrenaline (or epinephrine) and other similar substances. Adrenaline was found to contain a powerful 'pressor' substance, which raised blood pressure in experimental animals. The German neurologist Max Lewandowsky found that injecting the extract into cats produced retraction of the nictitating membrane (a protective membrane over the eyeball). He then demonstrated that it had the same effect on the eyeball if applied locally after cutting the nervous connection in the brain, showing that it acts directly on the muscle rather than on the nerve.

The English physiologist John Langley showed in 1901 that electric stimulation of the sympathetic nerves has the same effect as injecting the 'pressor principle'. The results of Lewandowsky and Langley taken together clearly suggested a chemical means of transporting the electrical impulse to the muscle. The English physiologist Thomas Renton Elliott suggested in 1905 that stimulation of the sympathetic nerves causes them to produce the pressor, which he named adrenaline, at their endings, and that this passes into the muscle and has a physiological effect. In the 1960s, Bernard Katz and Paul Fatt showed that receptors in muscle fibres are stimulated by the release of acetylcholine to open ion channels in the muscle membrane, producing an electrical current which causes the muscle to contract.

From 'vagus stuff' to neurotransmitter

The conclusive experiment came in 1921, and was carried out by the German physiologist Otto Loewi. Loewi claimed that the idea for the experiment came to him in a dream. He woke in the night and wrote it down, but in the morning he could not make out what he had written. When he had the same dream again, he got out of bed and went to the lab in the middle of the night to try it immediately. His experiment demonstrated how bioelectricity and chemicals work together in the transmission of nerve impulses. As was so often the case, it involved some unfortunate frogs.

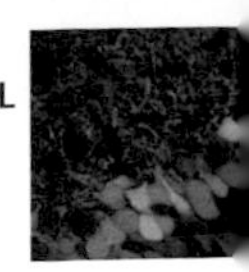

The sympathetic nervous system controls automatic defensive responses in the body, preparing this cat for 'fight or flight'.

Loewi knew that stimulating the vagus nerve caused the heart rate of a frog to slow down, and stimulating the accelerator (sympathetic) nerve caused it to speed up. He hypothesized that these actions made the nerve release a chemical that would cause a change in heart rate; stimulating the vagus nerve would produce a chemical that slowed the heart rate, while stimulating the accelerator nerve would produce a chemical that increased it. Loewi stimulated the vagus nerve of a beating frog heart until the heart slowed down. Then he collected liquid from around the heart and added it to a second heart in which the connections with the vagus and accelerator nerves had been cut. The second heart slowed down, confirming his hypothesis. Loewi named the chemical 'vagusstoff' (literally, 'vagus stuff'). It was later identified as acetylcholine, the principal neurotransmitter of the sympathetic nervous system.

Loewi was lucky that his experiment worked at all. By good fortune, he had chosen a species of frog with both excitatory and inhibitory fibres in the vagus nerve, but had carried out his experiment in winter when the inhibitory fibres predominate. His laboratory was cold, so the action of the enzyme which breaks down acetylcholine was slow, and enough acetylcholine remained to have an effect on the second heart. If he had conducted the experiment in summer, it might not have worked.

Nerve to nerve

The mechanism which enables a neuron to transmit a signal to a muscle also passes signals from one neuron to another across a synapse. Although it was easiest to discover

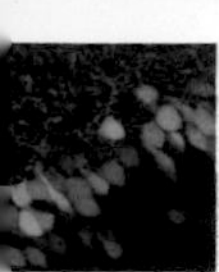

KILL OR CURE

Many chemicals can interrupt the transmission of nerve impulses at the point where it becomes chemical rather than bioelectric. The synapses are the most vulnerable place in the transmission circuit. Nerve agents such as Sarin are neurotoxins; they are used illegally as weapons (they are banned under international treaty). Sarin works by blocking the action of the enzyme acetylcholinesterase; this disrupts the transmission of nerve impulses across the synaptic gap. Acetylcholinesterase is responsible for breaking down acetylcholine once it has done its job of transmitting a nerve impulse to a muscle. The effect of blocking the action of this enzyme is that acetylcholine builds up and the muscle is not able to stop contracting. Death eventually results from asphyxiation. This is the opposite effect to curare, which works by preventing contraction. Painkillers, on the other hand, work to our advantage by blocking pain signals to or in the brain. Aspirin, for example, works by blocking the production of prostaglandins, which send pain signals to the brain, reducing our awareness of being hurt.

A terrorist attack using Sarin gas on the underground in Tokyo, Japan, in 1995 left 12 people dead.

A CONNECTION, AFTER ALL

In the 21st century it was discovered that some synapses in the brain *do* have direct electrical connections after all. The neurons act as a single unit and there is no neurotransmitter release. And sometimes more than one neurotransmitter may be released at a presynaptic terminal. This makes communication between neurons more complex and flexible than previously supposed.

the action of acetylcholine in the neuron–muscle junction, it was later demonstrated that neurotransmitters operate between neurons, too. Many physiologists, though, believed that bioelectricity was the only force at work and rejected any chemical mediation between neurons.

Squids to the rescue

In 1939, English physiologists Alan Hodgkin and Andrew Huxley tested the model of 'action potential' using the giant

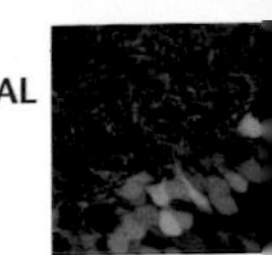

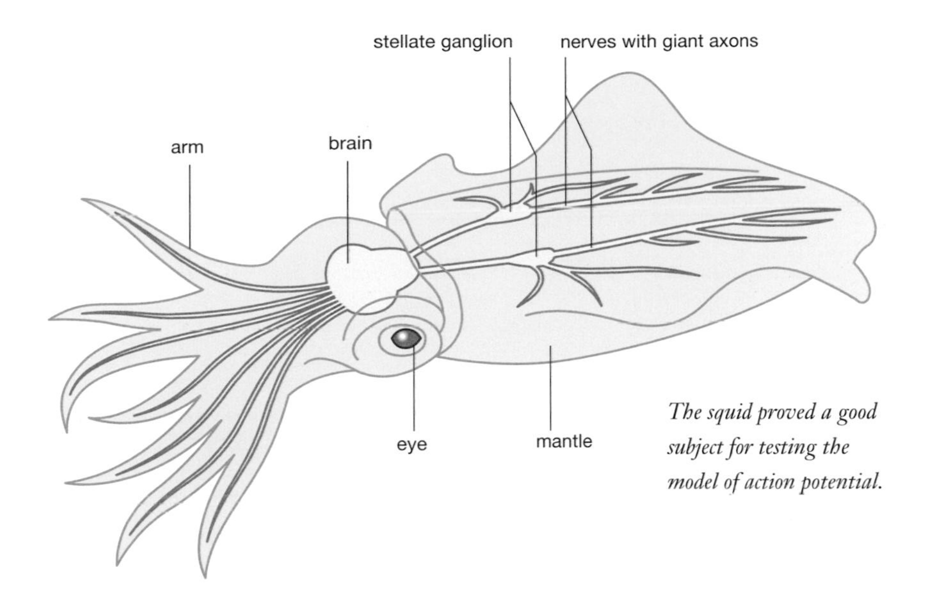

The squid proved a good subject for testing the model of action potential.

NEGATIVE AND POSITIVE IONS

Unlike the movement of electricity through a wire, a nerve signal or 'impulse' is chemical in origin and based on the passage of ions through the cell membrane. Ions are molecules that carry an electrical charge, either negative or positive. When a neuron is resting (not carrying any impulse) there is a difference in the concentration of negative and positive ions inside and outside the neuron. At rest, the neuron has an overall positive charge and the fluid outside the cell has a negative charge, so the membrane is polarized. This is called the 'resting potential' of the neuron.

The movement of ions is managed by proteins in the cell membrane, called ion 'channels', 'gates' or 'pumps'. Ions can only pass through the membrane at these points. Movement of ions in and out of a neuron via gates and pumps changes the polarity of the cell, from 'resting potential' to 'action potential'. The idea that a selectively permeable membrane surrounds a neuron and that this produces the resting and action potentials was first suggested by the German physiologist Julius Bernstein in 1902.

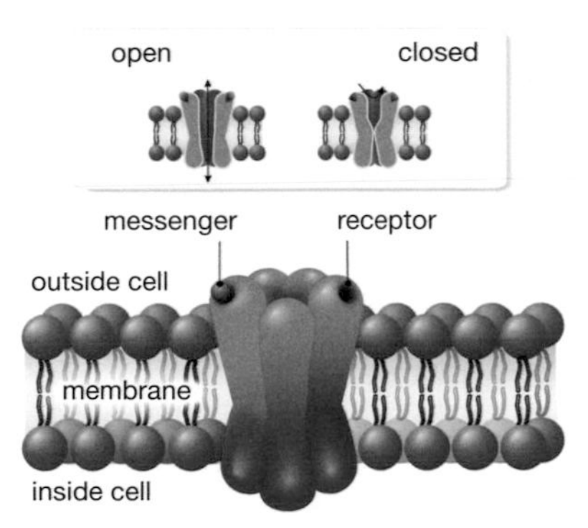

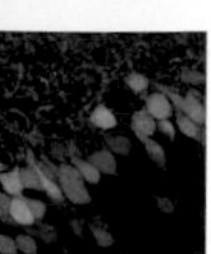

axons found in squid (see page 107). The squid has large neurons, with an axon around 1mm in diameter, which means it is visible to the naked eye and easy to manipulate.

The principle for moving an electric current along an axon had been outlined in basic terms by Galvani at the end of the 18th century. He described a theory of 'electrical excitation' in which the resting tissue is in a state of 'disequilibrium' – that is, ready to respond to external stimuli by generating electrical signals. Galvani compared the mechanism he envisaged to a Leyden jar, a device that stores static electricity between its inner and outer layers. In his model, 'animal electricity' is the result of accumulating positive and negative charges on the external and internal surfaces of the muscle or nerve fibre. He proposed that water-filled channels penetrating the surface of the fibres allow the flow of charges in and out, producing electrical excitability. Referring again to the analogous Leyden jar, he went on to suggest an insulating covering, with holes to allow the electric charge through at some points.

Huxley and Hodgkin found the mechanism described by Galvani at work in the squid's giant axons. They sought

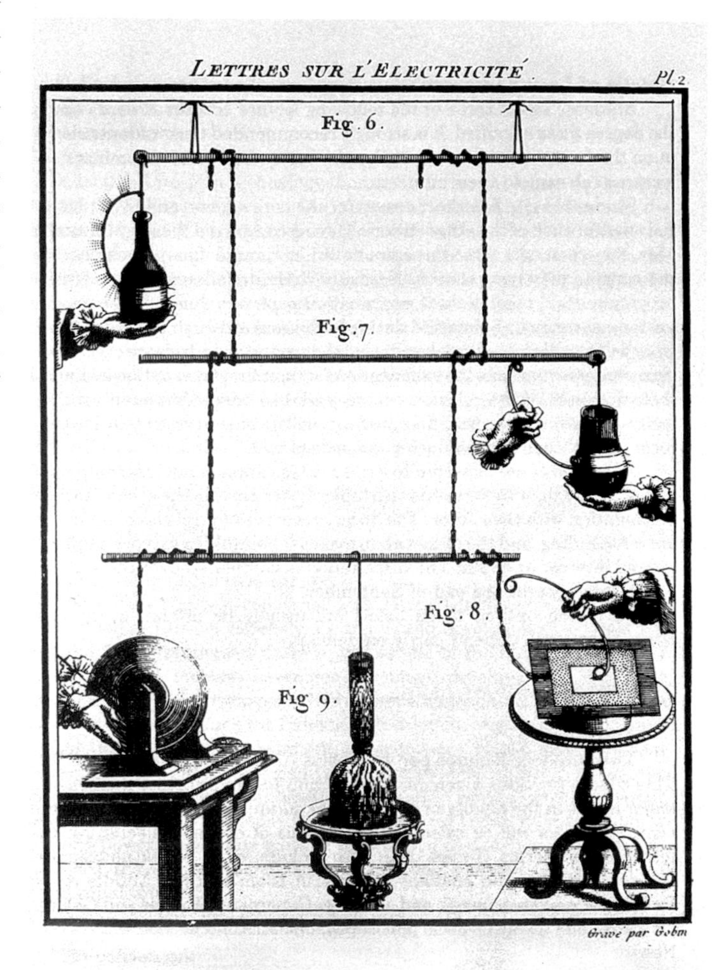

A Leyden jar stores static electricity produced by turning a glass sphere (lower left). The electricity can then be used to produce a spark or shock by completing the circuit between the jar's inside and outside conducting layers.

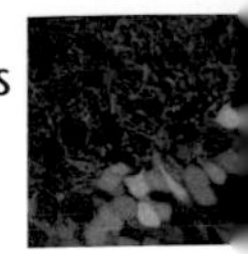

to discover the voltages involved and the type of ions that were moving through the membrane to produce the action potential. Because the squid axon is so large, they were able to thread a wire through it and attach it to electrodes. They could then maintain and measure the voltage across the cell membrane. This device is called a voltage clamp, now a basic tool in electrophysiological research.

For a neuron to produce an electrical impulse, the voltage inside and outside the neuron must be different. The neuron then produces a gradient of charged ions that enter or leave the axon, producing rapid changes in voltage – the action potential. The voltage clamp reads the voltage in the axon and supplies sufficient current to keep it at the level the researcher has chosen. As the clamp constantly adjusts the voltage to make up for the effect of ion transfer across the membrane, recording the adjustments needed shows the action of the ions (an equal but opposite current to that supplied by the voltage clamp). This allowed Hodgkin and Huxley to measure the changing voltage of the action potential in the cell. They published their results in 1952.

In 1961, Peter Baker, Alan Hodgkin and Trevor Shaw experimented by replacing the cytoplasm in the squid axon with a variety of other ionic solutions. They discovered that the ions which pass through the membrane to create the action potential are sodium (Na^+) and potassium (K^+). When the neuron is at rest, the membrane is mainly permeable to potassium, but when the action potential is fired, the membrane becomes more permeable to sodium.

SO NERVES WERE HOLLOW AFTER ALL?

After all that effort to prove that nerves are not hollow and don't allow the flow of animal spirits, it turns out that at the most basic level they are made of neurons which actually do contain fluid. But the fluid doesn't flow from one place to another – it stays in the neuron. What flows is the electrical current within the fluid in the neuron.

Connections and connectomes

A basic understanding of the mechanisms of the nervous system was established by the end of the 20th century. It was clear that impulses are carried along neurons as an action potential, produced by ions carrying an electrical charge. The gap of the synapse is 'jumped' by chemicals released across the

BRAIN SOUP

There can be millions of neurons in a tiny amount of human brain tissue, but how many are there in the entire brain? Until 2012, no one knew for sure. Then Brazilian neuroscientist Suzana Herculano-Houzel took samples from the brains of men who had died from non-neurological causes and turned each one into mush. She then counted the neurons in a set volume of this 'brain soup' and from this estimated the total number of neurons in the whole brain. She found the average to be 86 billion.

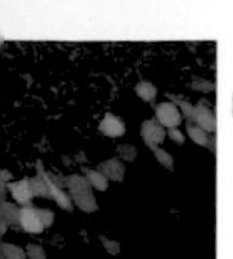

ALZHEIMER'S DISEASE

Alzheimer's disease causes dementia and behaviour changes. Autopsy reveals a build-up of plaque in the brain made of amyloid-beta protein, a natural byproduct of brain activity. The protein is usually cleared away, but in people with Alzheimer's the clear-up mechanism does not work properly and sticky plaque forms which prevents signals crossing synapses. In 2016, tests in mice with Alzheimer's showed that treatment with the brain enzyme BACE1 prevents strands of amyloid-beta clumping together and stops the development of plaque. This could possibly be developed into a treatment for Alzheimer's.

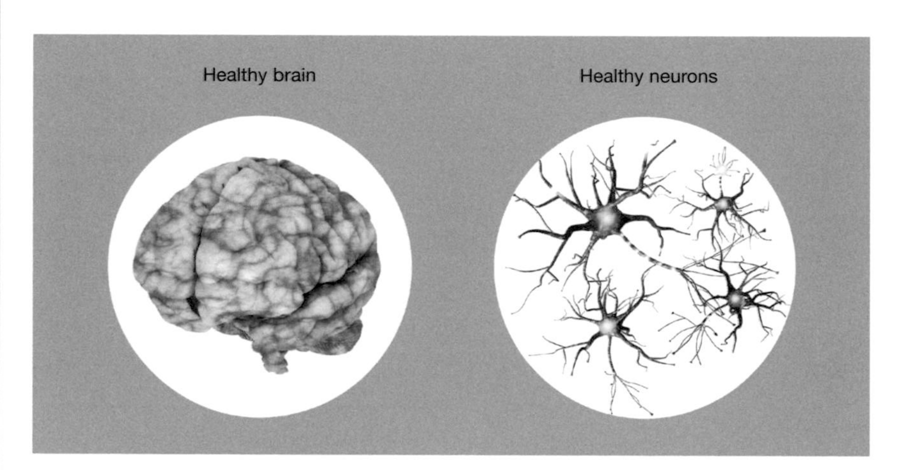

Alzheimer's disease

Diseased neurons

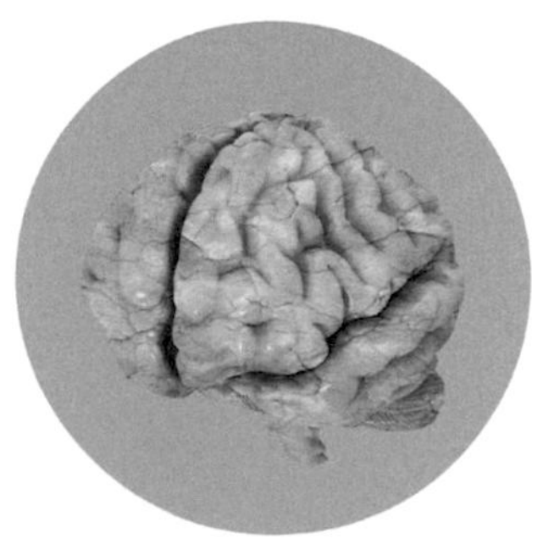

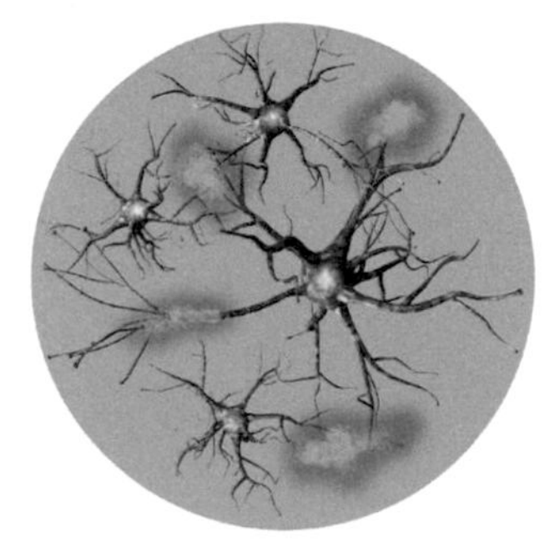

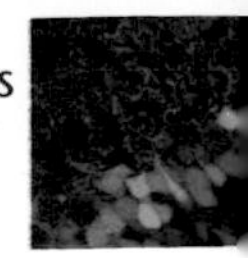

The movement of water traced in a healthy brain revealed by MRI scan shows connections within the brain.

synaptic gap. This kick-starts the electrical signal in the next neuron or prompts an adjacent muscle to contract. It now remained to investigate the incredible network of connections and what they do.

The peripheral nervous system is the easiest to explore as it's a relatively simple task to trace neurons to their terminations in the skin, in muscles, in an internal or sensory organ. But most of the neurons in the human body are inside the brain, forming part of the central nervous system. Working out what they do and how they are connected – each might have as many as 7,000 connections – is a huge task. Yet it is one that the Human Connectome Project (HCP) has set itself (a connectome is essentially a wiring diagram for a brain or organism). The HCP is a consortium of researchers based in Washington University, the University of Minnesota and Oxford University. It aims to map the brains of 1,200 healthy adults using cutting-edge neuroimaging techniques, releasing the data in quarterly updates (starting in 2013). It is the neuroscience equivalent of the Human Genome Project – ambitious, daunting, but offering immensely valuable insights into the workings of both healthy and unhealthy brains.

CHAPTER 6

Senses and SENSIBILITY

'Sensation consists in the sensorium receiving through the medium of its nerves, as a result of the action of an external cause, a knowledge of certain qualities or conditions not of external bodies but of the nerves themselves.'

Johannes Müller, 1840

While the motor system's principal task is to cause muscles to contract, thereby producing movement, the primary job of the sensory system is to detect stimuli from outside and within the body. These stimuli, which include light, sound, smell, touch and taste, are processed by the brain to enable us to interact with our surroundings. The results of the processing might activate the motor system or they may remain internal to the brain, producing memory, understanding or emotion.

A ride such as this provides a multitude of sensory signals to the brain, producing both physical and emotional responses.

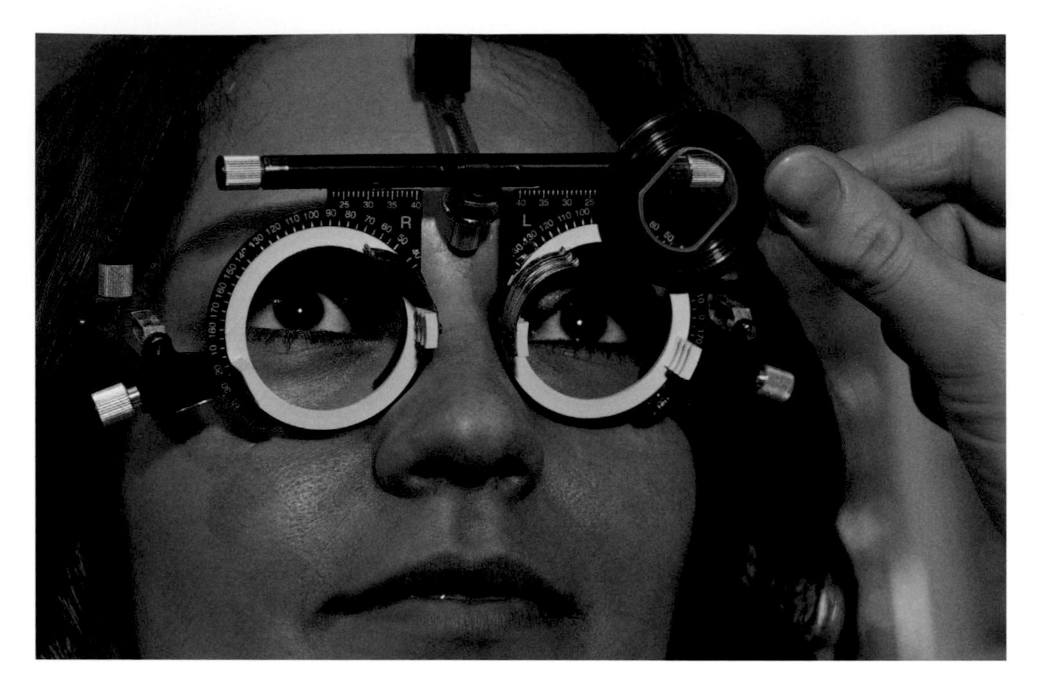

An optometrist can correct faults in the sensory organ – the eye – using lenses or surgery, but addressing vision problems that originate in the nerves or brain is much more difficult.

Dividing the work

Even the ancients distinguished between sensory organs that acquire data and the 'sensorium', be it brain or heart, which processes the data into a sense experience of which we are conscious. The mechanism can be broken down into the reception of a stimulus, transfer of the stimulus to the sensorium, and processing of the stimulus.

Early work on the senses focused on how sensory perceptions work mechanically. Later, scientists began to wonder about how much the senses have in common, and how they differ.

Inside and outside

The external senses – vision, hearing, touch, taste and smell – are all concerned with the relationship of our bodies to the outside world. This means that in discovering how they work, we not only need information about the way our nerves and brains work, but also some knowledge of how matter is constructed and of phenomena such as light and sound. Before we developed a good understanding of topics such as optics, audio and chemistry, there was really no hope of knowing how our senses work.

The mind's eye?

The first sense to be explored extensively was sight. In part, this may be because vision is so very important to us, but it may also be because it is a sense we can easily shut off, simply by closing our eyes.

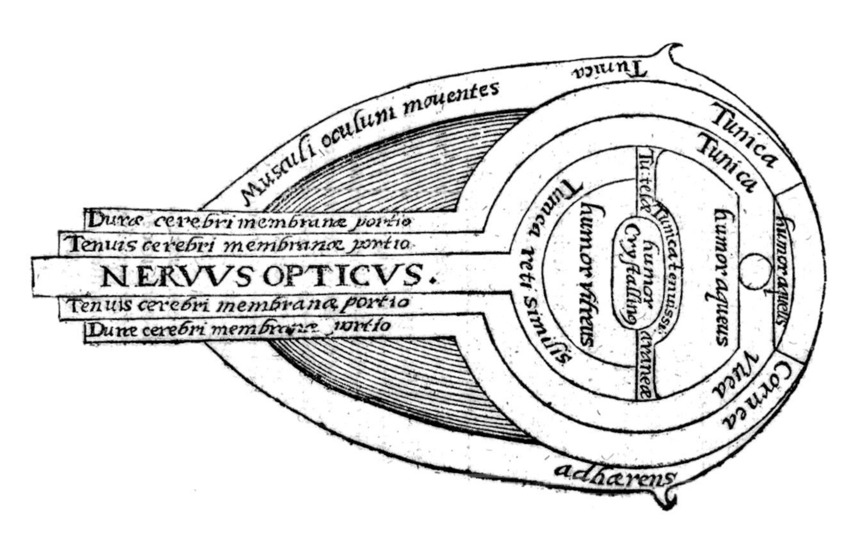

Alhazen ibn al-Haytham showed the optic nerve leading from the back of the eye, originating in the innermost of the 'tunics' thought to make up the outer structure of the eyeball.

Carrying impressions: in and out

Perhaps it is this ability to 'turn off' vision that first led to the idea that the eye sends out a beam of some kind that seizes an impression of objects. This is known as the 'emission' or 'extramission theory of vision' and originates with Alcmaeon and Empedocles in the 5th century BC. The idea was explained by Democritus, who said that objects emit something into the air, making a kind of imprint on it. The Greek philosopher Theophrastus (*c.*371–*c.*287BC) considered the air between an object and the eye as though it were solid, so an impression created on the air is pushed against the eye like a stamp pushed against wax, facilitating vision.

One clear objection to the extramission model is that if light comes from the eyes to capture an image of the objects around, then we should be able to see in the dark. Empedocles had an answer to that: there is some sort of interaction between the rays that come from the eyes and some external source of light, such as the sun's rays. After Empedocles made his views known, Plato, Ptolemy and Galen all endorsed the extramission theory, giving it considerable weight in Europe and the Arab world until the 18th century.

Aristotle, on the other hand, preferred the intromission theory – that light is carried to the eye. His view was supported by some of the Arab scholars of the Middle Ages

PERCEPTION AND MATTER

The Greek philosopher Democritus (*c.*460–370BC) was the first to propose an entirely physical means for perception. He described sense impressions in terms of the senses receiving 'atoms' emanating from matter and relaying them to the brain.

who wrote extensively on optics and vision. In the early 10th century, al-Razi wrote of how the pupil contracts and dilates, and in the 11th century al-Haytham remarked that strong light can injure the eye. Ibn Sina also argued in support of intromission. But these arguments were not enough to oust extramission as the favoured model.

Optics and optic nerves

Galen considered the retina and optic nerve to be extensions of the brain. He proposed that 'visual spirits' travel along the optic nerve and across the eye to the lens (the 'cystalline humour'), which he considered to be the primary element in the visual system. At the lens, he believed, the spirits mixed with light from outside the eye, picked up a visual impression, and then returned it along the optic nerve to the brain. The same system was described by Master Nicolaus of Salerno nearly 1,000 years later.

The tide turns

Leonardo da Vinci, originally a proponent of extramission, changed his mind at some point in the 1480s or 1490s. In 1583, the Swiss physician Felix Platter challenged the idea that the lens was the most important part of the eye in terms of receiving visual information and stated instead that the optic nerve is the primary organ of vision. This opened the way for the consideration of the retina's importance.

From light to sight

Descartes was one of the first people to try to explain perception in terms of the physical mechanics of light and the body. He effectively separated vision into two parts, one of which was mechanistic while the other was carried out by the brain reading the visual information and constructing vision. This fitted Descartes' dualist model of body and soul being separate, but communicating. It also corresponds

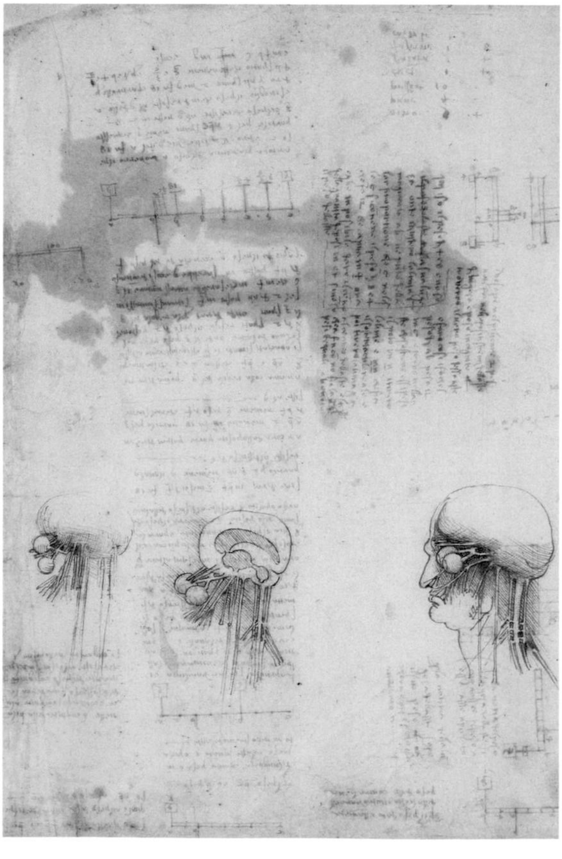

Leonardo da Vinci's drawing of the eyes and brain shows the optic nerve entering the front of the brain.

HOW TO DISSECT AN EYE

Leonardo advised removing the eye from a cadaver, placing it in egg white and boiling it so that it was fixed in the cooked egg. It was then easier to slice it for examination (without a microscope, of course).

roughly to the way we still think of vision. We now know that colour is the result of a surface reflecting light of a particular wavelength and the excitation of cells in the retina which are sensitive to that particular wavelength. This is interpreted by the brain as the colour that we see. Although he had the physics wrong, Descartes correctly had the brain construct the experience of colour from sensory input.

Learning to see?

Descartes endowed the brain with an innate capacity to read the information coming from the eyes, in order to see and interpret what was seen. In 1688, long after Descartes' death, the Irish natural philosopher William Molyneux wrote a letter to John Locke in which he asked whether if a man blind since birth was suddenly able to see he would be able to distinguish a cube from a sphere by vision alone. The issue, which became known as the Molyneux Problem, sparked a debate about whether seeing is innate or learned. Molyneux was of the opinion it is learned, and that the suddenly sighted man would not be able to distinguish the cube from the sphere. The man would have a 'schema' – a model of the world – developed through touch, but there would be no automatic way to map what he could see to that schema. Locke agreed with this view.

LASTING EYE-BEAMS

The English poet, John Donne, could still confidently use the idea of eye-beams in the early 17th century:

'Our eye-beams twisted, and did thread
Our eyes upon one double string.'

John Donne (1573–1631), 'The Extasie'

Descartes shows visual information being transferred from the eyes to the pineal gland, which he considered the seat of the soul.

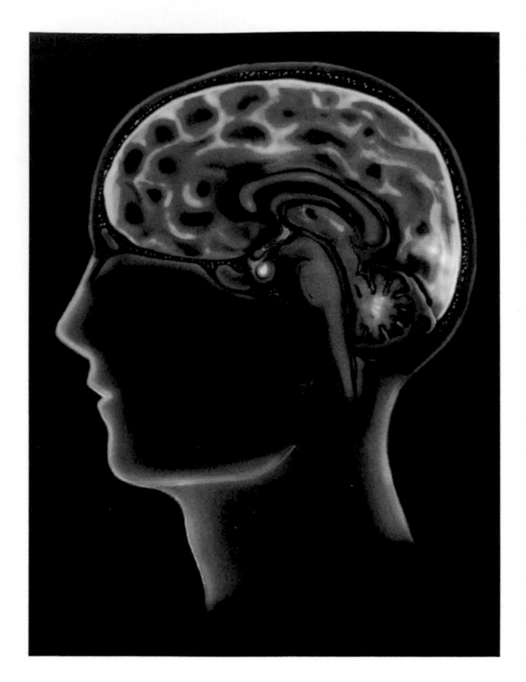

The occipital lobe at the back of the brain is the area involved with vision.

It turns out that both men were right. In 2006, tests in India and the USA with individuals blind since birth, but enabled to see after surgery, revealed that the subjects were not innately able to correlate between visual data and a schema built up by touch, but they could 'learn to see' until at least late childhood. Tested some time later, subjects had 80–90 per cent of the recognition skills of a person who had been born sighted.

Back to rays

In 1604, the astronomer Johannes Kepler (1571–1630) described how the lens focuses an image on the retina and it is this that conveys visual information to the brain. The principal difficulty with Kepler's theory was that an image captured in this way would be inverted by the lens, but we don't see the world upside down. Yet it turns out to be true – the image is inverted even though we don't see it that way. This was even demonstrated by Descartes using the eyeball of an ox. An image projected through an ox's eye is, indeed, upside down. The retina receives an image that shows everything upside down, but our brain inverts it for us. Kepler suggested as much while claiming that it was not his job to worry about how it happened, but that he suspected the 'activity of the Soul' turned the image the right way up.

SEEING UPSIDE DOWN

In the 1890s, the psychologist George M. Stratton experimented with vision by wearing special glasses that obscured one eye or inverted or reversed the image he saw. He found that while he saw the world inverted or reversed for a few days, his brain invariably adjusted to the change and righted the image. When he stopped wearing the glasses, he experienced the same inversion until his brain corrected the image again.

Newton's eyes

Isaac Newton (1642–1727) was a polymath whose disciplines spanned physics, astronomy, optics and alchemy. His work with light led him to experiment with his own vision, provoking his optic nerve to deliver different images under different conditions. He was not squeamish, and was happy to put pressure on his eye with

a finger, a needle and a brass plate in order to distort the eyeball and discover the effects: 'I tooke a bodkine & put it betwixt my eye & bone as neare to [the] backside of my eye as I could: & pressing my eye [with the] end of it . . . there appeared severall white darke & coloured circles.'

Newton noted that if he looked directly at the sun and then at a sheet of white paper, he could see a circle where there was none. Then he discovered he could replicate this merely by imagining he had looked at the sun. He concluded that vision could be manipulated, or the optic nerve tricked, into delivering something that is not really there. This suggested that vision is not entirely a mechanical process after all, as how could fantasy make him see something if it were? He concluded that what we see is affected by the nerves – therefore he saw strange effects if he compressed his eye – and he also decided that there could be something of 'spirit' in perception after all. This spoke against Descartes' image of the body as a machine that could be fully explained in mechanistic terms.

Between them, Molyneux and Newton had addressed the two different aspects of vision. Newton concentrated on the physical, the optics. Molyneux was more

Isaac Newton was fearless to the point of foolishness in his experimentation on his own vision.

NOT PERSUADED?

Newton and Locke were instrumental in explaining and promoting the intromission model in the 17th century. But they have not been entirely successful: a study in 2002 found that around 50 per cent of American adults still believe that something emanating from the eye enables vision.

interested in what the brain did with the information gathered and passed on by the eyes – the 'activity of the Soul'.

Broken rainbow

The English physicist Thomas Young, famous for establishing the wave theory of light, proposed in 1802 that there might be three types of receptors in the eye for detecting the three primary colours. From these, all other colours could be perceived by mixing the primaries in different proportions.

Improvements in microscope technology proved him right. In 1838, Johannes Müller distinguished in the retina a layer of shapes that looked like rods packed closely together, and in 1852 Rudolph von Kölliker identified separate rods and cones. The following decade, Max Schultze proposed that two different types of receptor deal with different aspects of vision: the rods enable night vision (in shades of grey) and the cones enable colour vision in daylight. The exact mechanism became clear in the 20th century, with the discovery that different chemicals within receptor cells respond to different wavelengths of light.

HEAR, HEAR

Just as understanding vision was tied in with developments in optics, so understanding hearing was tied in with understanding how sound works. Until Robert Boyle discovered that sound waves have to travel through a medium (such as water or air) in 1660, there was little hope of making any progress. It was not until the mid-1700s that more detailed understanding of sound developed with Daniel Bernoulli's work on vibrations and frequency.

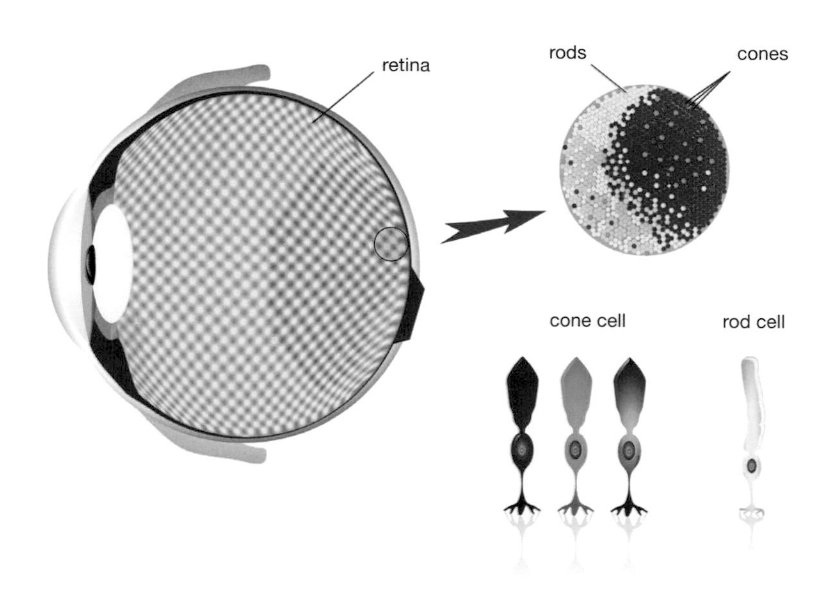

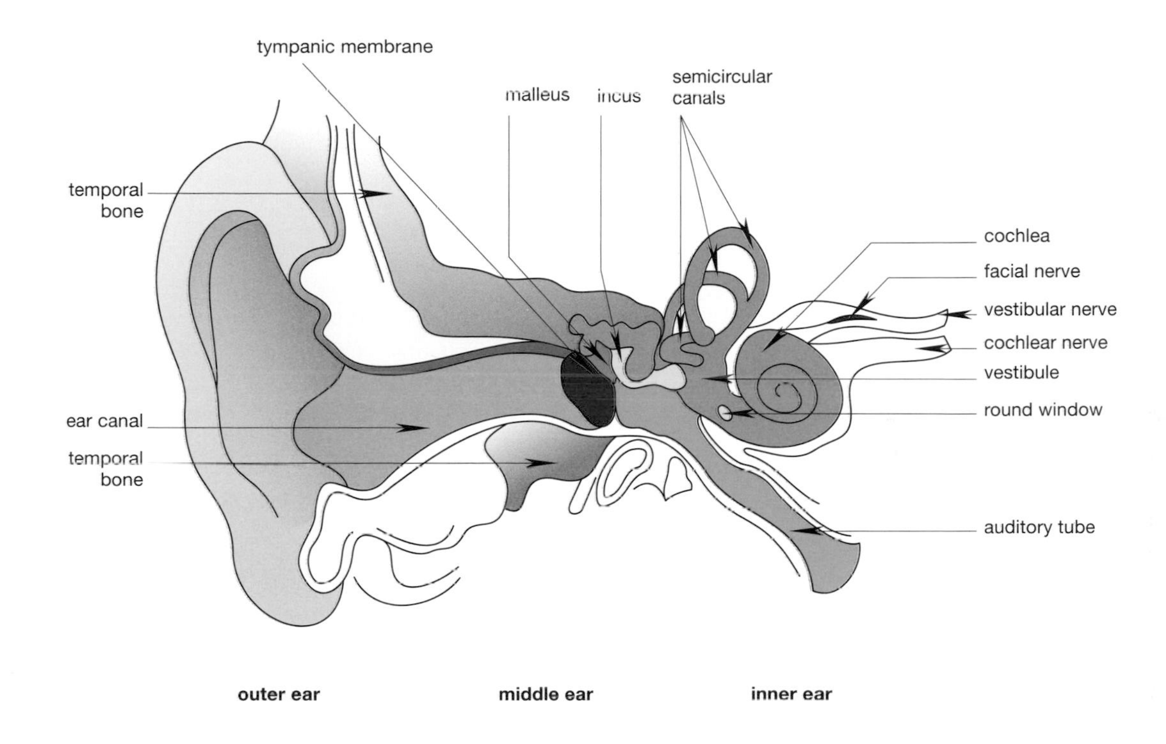

The anatomy of the external and internal parts of the ear.

Inner air and inner ear

The Ancient Greeks were almost unanimous in believing that there is some kind of purified air inside the ear, which arrives as the foetus develops *in utero*. Vibrations in the external air were thought to be communicated through the ear to this inner air, which passed a sound impression to the brain. Belief in this special 'inner air' persisted into the 18th century, even as the anatomy of the outer, middle and inner ear was uncovered.

COCHLEAR IMPLANTS

People who cannot hear because the hairs in the cochlea are damaged can be helped by a cochlear implant, invented in 1982. A unit on the outside of the ear digitally codes sound and transmits it to a device implanted in the inner ear. This bypasses the process of hearing by directly stimulating the nerves in the ear which then send signals to the brain.

Hair, not air

The middle ear is indeed filled with air, but it's not a special kind of air and serves only to pass sound waves on to the inner ear. The inner ear is a spiral tube called the cochlea, which is filled with fluid and has a membrane covered in tiny hairs. Vibrations from the middle ear are communicated through the fluid to the hairs. The hairs are associated with receptors which send a signal when stimulated by a hair moving. Hairs in different areas of the membrane are activated by different sound frequencies (and interpreted as tones).

Italian anatomist Alfonso Corti discovered the tiny hairs in the cochlea in 1851, and Swedish neuro-anatomist Gustaf Retzius observed the nerve endings near them, but it was not until 1937 that the nerves of the ear were properly understood. Spanish neuroscientist Rafael de Nó showed that each hair was served by one or two nerve fibres, and each nerve fibre split to serve only a few hairs. The movement of a hair triggers the release of a neurotransmitter which, in turn, triggers the cochlear nerve to send a signal to the brain.

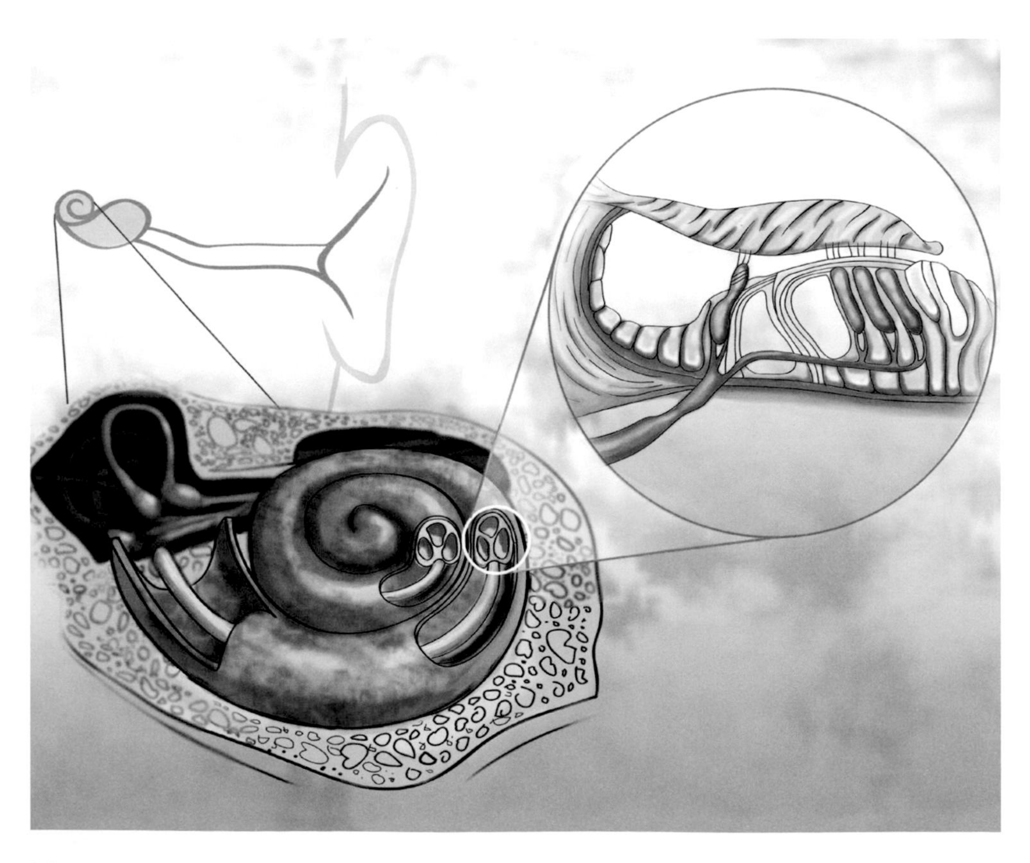

The arrangement of hairs and nerves in the cochlea.

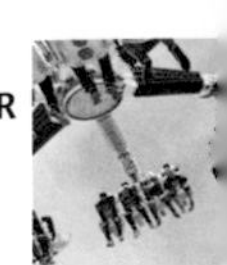

THE MCGURK EFFECT

Usually, our brains put together information from our senses and successfully interpret a composite stimulus, such as something we can both see and hear. But if the message from two senses conflicts, one has to be favoured. Speech perception is multi-modal – it takes both sound information and visual information (if we can see the lips of the person speaking).

In an effect discovered in 1976 by British psychologist Harry McGurk, if we see a mouth forming one sound but are played the audio of a different sound, then the sound we are conscious of is the one that matches the lip movements, not the one we hear. This can be demonstrated using the sound 'ba', first played matching the movement of the speaker's mouth and then with the mouth forming 'fa'. The subject hears this as 'fa', not 'ba', as vision has taken priority over hearing; in some cases, the subject may hear a third sound, as a compromise between the conflicting signals.

Hear and here

During the 19th century two models of how the ear received sound were current. In one model, different locations within the ear were considered receptive to different sounds; in the other, there was no localization within the ear. This mirrored the contemporary debate about the localization of functions in the brain. Hermann von Helmholtz took specialization to extremes, suggesting that there were distinct receptors for every one of the 5,000 different tones he believed the human ear could distinguish.

Localization of hearing need not be as specific as Helmholtz's requirement for different receptors for every possible sound. The more common model had high and low frequency sounds received by different parts of the cochlea. Evidence for this theory was taken from examining people whose hearing had been damaged by repeated exposure to loud noises, particularly men who worked on the railways. It also relied on exposing animals (often guinea pigs) to loud noise close to their ears until eventually their hearing was damaged. It transpired that playing high-frequency noises led to damage to the basal area of the cochlea.

Researchers believed that different areas of the cochlea were associated with perception of sounds of different frequencies. But the damage discovered was not restricted to the areas they thought detected sounds of the frequencies used in the experiment. Although repeated exposure to high-frequency sounds readily caused damage, the same was not true of low-frequency sounds at high volume. It seemed that the perception of low-frequency sounds was not localized in the cochlea in the same way.

The 'telephone' system

An alternative to a localized theory was what William Rutherford called the 'telephone theory' of hearing. He suggested that the ear does not distinguish between sounds, but all hairs vibrate in response to

William Rutherford suggested a new model for hearing - one that could only have developed from the electromechanical developments of the 19th century.

all sounds. The vibrations are converted into nerve-vibrations that are carried to the brain by the auditory nerves. There, the brain interprets them and puts the various information about frequency and amplitude together to produce the 'sensation of sound'.

The tongue and taste

Taste is a different kind of sense in that a physical substance clearly is involved. The Ancient Greeks were generally agreed that taste results from tiny particles entering pores on the tongue and being transported to the organ responsible for processing sensory input, whether that was the heart or the brain. Democritus believed the shape of the atoms determined how they interact with the body: round, large atoms produce sweetness, large angular atoms produce astringency, and so on.

Aristotle had a slightly different model. He identified seven basic tastes: sweet, sour, bitter, salty, astringent, pungent and harsh. He believed the qualities of the taste, rather than atoms, were transferred to the tongue and carried by the blood to the heart (which he regarded as the control centre). Galen noted the need for the tongue to be moist for taste to function properly, so included the salivary glands in the machinery of taste. He tried to work out the innervation of the tongue, but was wrong regarding which nerves are responsible for moving the tongue and which are involved in the sense of taste.

Four basics

After many further attempts to name the fundamental tastes, they were finally whittled down to the basic four – sweet, sour, bitter and salty – by Maximilian von Vintschgau

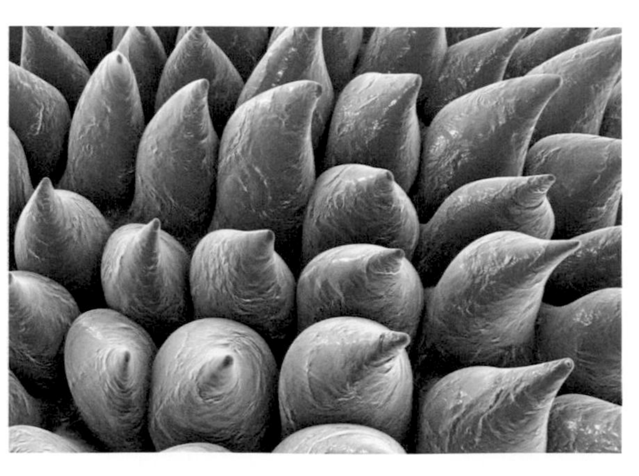

Papillae on the tongue, greatly magnified.

in 1880. Debate continued, though, with some people wanting more and some fewer basic tastes. The suggestion that there might be different processes for sensing the four basic tastes followed the discovery that anaesthetics such as cocaine affect the perception of tastes differently, with the ability to detect bitter tastes being lost first (straight after the loss of pain sensation). Work on discovering how different parts of the tongue are sensitive to different tastes began in the 1820s, but became more detailed in the 1890s and was quantified in the early 20th century.

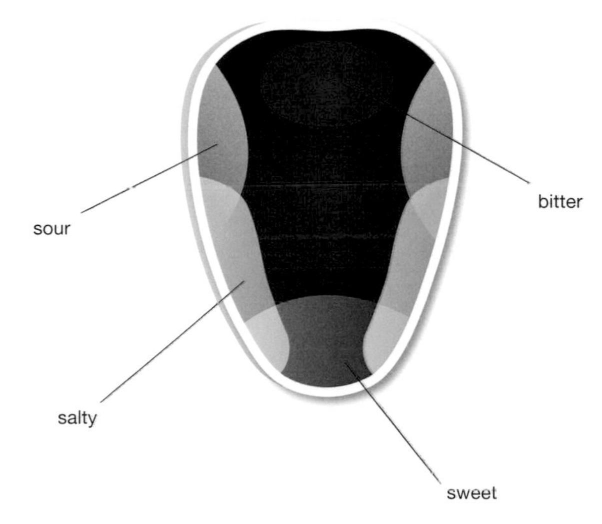

Different areas of the tongue are particularly sensitive to different types of taste.

It was clear from the 18th century that the tongue is not the only place capable of sensing taste. The physiologist Claude le Cat revealed in 1750 that he had studied two children devoid of tongues (one was born without a tongue and the other lost it after an infection), and both were able to distinguish tastes.

From taste buds to tastes

That the tongue is covered with papillae is obvious to the most casual observer. Albrecht von Haller suggested in 1747 that they might be the organs of taste, a verdict that Johannes Müller and Charles Bell both agreed with early in the following century. Bell demonstrated, by poking the papillae with a metal probe, that some of them detected touch and others detected taste (the taste of the metal in this case). But later studies by the Swedish physician Hjalmar Öhrwall (1851–1929), investigating whether papillae are specialized to particular tastes, suggested they are not. He found that most papillae can detect at least two different tastes, but clearly there is a finer organ for detecting taste.

Considerable work went into determining which areas of the tongue are most sensitive to different tastes. It became clear, too, that the sensitivity to taste changes with age: babies and young children have a better developed sense of taste than adults, and taste sensitivity drops off after about the age of 45.

The taste buds were first discovered in 1867 independently by researchers working with animals, and were soon found on the human tongue as well. They proved difficult to investigate. Estimates for the number of taste buds on each papilla varied widely, and the taste buds turned out to be too close together to allow experimenters to work on them individually.

The details of which nerves serve taste and where in the brain taste is processed were slow to unravel. The time-honoured process of examining lesions, or cutting nerves and documenting the effect, worked less well with taste than with the other senses. Few lesions seemed to have a clear effect on taste, and there were fewer experiences of taste auras among epileptic patients than other types of aura. (An aura is a sensory disturbance, such as seeing a light or experiencing a smell that has no external trigger.) Scottish neurologist David Ferrier associated taste with the frontal lobe, and this view was generally accepted until the mid-20th century. Then, the examination of patients with gunshot wounds and the electrical stimulation of the brains of conscious surgery patients revealed that taste is processed in the parietal lobe of the cortex.

Sensitive skin

From the Ancient Greeks onwards, people disagreed about whether the skin provides one sense or several. Feeling heat is not the same as feeling a tickle or the prick of a pin, so are they in any meaningful way the same thing? The question of whether and how to subdivide touch was tackled by, among others, Ibn Sina in the 11th century, Albertus Magnus in the 13th, Francis Bacon in the 16th/17th and Immanuel Kant in the 18th.

Types of touch

Skin is our largest sensory organ. It covers the entire body and is able to detect several types of stimuli, now separated into heat, cold, touch and pain.

Touches such as tickles and caresses are detected by specific receptors in the skin, quite separate from those that detect other stimuli, such as heat or pain.

In the 4th century BC, Aristotle distinguished between paired qualities of touch, such as hard–soft and hot–cold. Galen considered that discerning these qualities was a learned response based on previous experience, therefore not quite so fundamentally different at the point of receiving the stimulus. He believed that information from the peripheral nerves travelled to the brain, where it was then interpreted to determine its qualities.

Anatomists from Ibn Sina onwards tried to divide up the aspects of what the skin can sense, but did not necessarily assume that they were sensed differently. In 1844, the Polish scientist, Ludwig Natanson (1822–71) suggested that the sense of touch can be divided into three parts, each with its own type of receptor organ: temperature, touch and tickle. Pain, he thought, was the result of activating all three at once. He based this

theory on his observation that as a limb 'goes to sleep', sensitivity to these different stimuli is lost separately and in the same sequence. Hermann von Helmholtz developed the idea of sensory modalities, dividing the skin's sensory reception into different areas which are then perceived along a continuum. In the late 19th century, the Austrian-born physiologist Max von Frey (1852–1932) made popular the view that there are essentially four different cutaneous

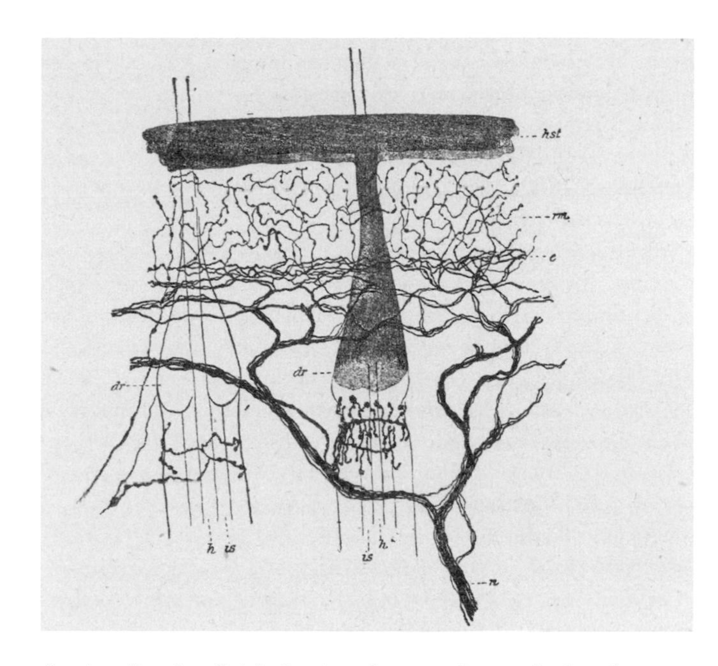

Santiago Ramón y Cajal's drawing of nerve endings in the skin of a rat.

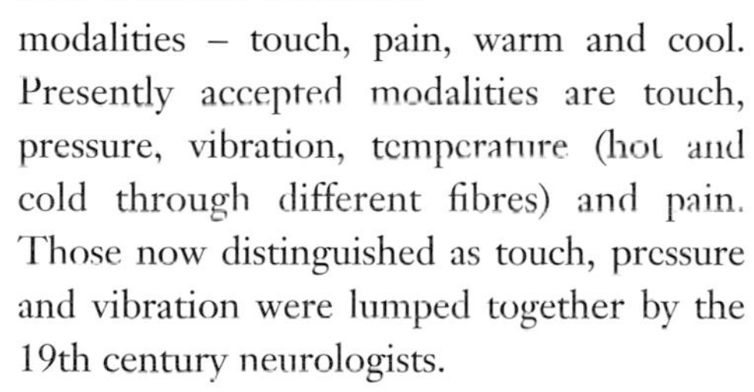

modalities – touch, pain, warm and cool. Presently accepted modalities are touch, pressure, vibration, temperature (hot and cold through different fibres) and pain. Those now distinguished as touch, pressure and vibration were lumped together by the 19th century neurologists.

Nerve endings were first identified by Abraham Vater in 1741 and rediscovered by Filippo Pacini in 1831. Thereafter, different types of receptor were described and named between 1848 and the 1930s. Initial studies concentrated only on description.

Just noticeable

The idea that the different types of sensation might be detected by different receptors was first proposed and explored by the German physician Ernst Weber (1795–1878) from the 1830s. He envisaged the skin as a mosaic of tiny areas each served by a nerve and set out to research how close together they are – therefore, how sensitive the skin is. Weber was working before the idea of modalities or the discovery of different nerve endings in the skin, and he believed the sense of touch included temperature, pressure and position and that the aspects did not operate independently. As an example of the last point, he cited evidence that pressure from a cold object seems greater than pressure from a warm object of the same weight.

Weber is best known for his work on the 'just noticeable difference' (JND); this is the threshold at which we can tell sensations apart. He began with the ability to sense differences in weight (or pressure), and discovered that a percentage change

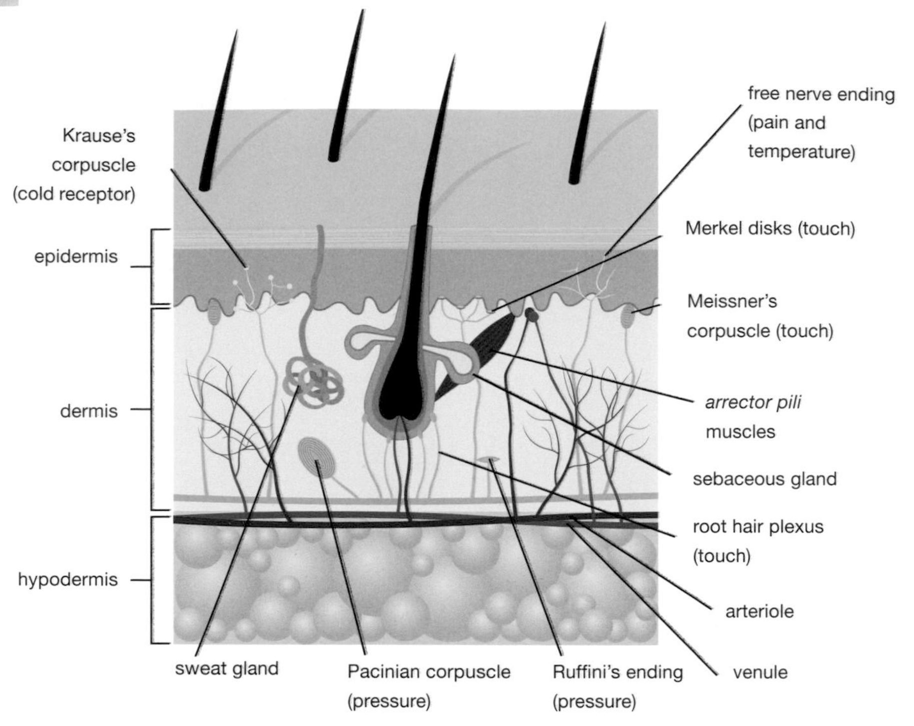

Different types of sensory receptor found in the skin.

in stimulus is needed for us to notice a difference, although the exact threshold varies between individuals. For example, a person may detect a difference between a weight of 30g and one of 31g, but a weight of 60g would need to increase to 62g before it was perceived as being noticeably different. He found thresholds for several different types of perception, and formulated his findings into Weber's law.

Just a little prick – or two

One of Weber's tests for sensitivity involved pressing the points of a pair of dividers against the skin to determine when a subject could distinguish between one and two sensations. If the points are very close together, they are felt as a single stimulus. He found that the ability to discern two points varies between different regions of the body and diminishes with fatigue. Weber also worked on our ability to localize sensations, asking blindfolded subjects to indicate the exact position on their skin that had been touched.

Matching nerve endings to sensations

The different types of nerve ending were discovered by various anatomists throughout the 19th century, including Wilhelm Krause (Krause's end-bulbs or corpuscles), Georg Meissner (Meissner's corpuscles), Friedrich Merkel (Merkel cells), Filippo Pacini (Pacinian corpuscles), Angelo Ruffini (Ruffini's corpuscles) and Rudolf Wagner

NERVE ENDINGS IN THE SKIN

Neurology now recognizes six types of cutaneous receptors for mechanical stimuli and four further types of receptor. There is overlap in their functioning, and between them they enable sensations of: touch, pressure, vibration, heat, cold and pain. In 1831, Filippo Pacini found receptors which detect pressure and vibrations. Georg Meissner and Rudolf Wagner found receptors sensitive to light touches in 1852, and Wilhelm Krause identified a receptor sensitive to gentle vibrations in 1860. Magnus Blix found temperature-sensitive receptors in 1882. Max von Frey found spots specifically connected with pain in 1896.

(a type of nerve ganglia), but it was not immediately clear what they all did. Deciding which type of nerve ending should be associated with each modality was initially a hit-and-miss affair, not based on rigorous anatomical evidence. Von Frey proposed that touch is linked to hair receptors and Meissner corpuscles, pain to free nerve endings, and warm and cold to Ruffini corpuscles and Krause end-bulbs respectively. He did not have particularly good reasons for these assignments, choosing to associate free nerve endings with pain simply because there are a lot of them.

There were two experimental approaches to working out exactly what each type of receptor does. One was to cut out some receptors and see

'According to the different types of apparatus with which we provide terminations we can send [through wires] telegraph despatches, ring bells, explode mines, decompose water, move magnets, magnetize iron, develop light and so on. So with the nerves the condition of excitement which can be produced in them and is conducted in them is, so far as can be recognized in isolated fibres of a nerve, everywhere the same, but when it is brought to various parts of the brain, or the body, it produces motion, secretion of glands . . . sensations of light, hearing, and so forth.'

Hermann von Helmholtz, 1863

A STUDY IN NERVES

In the early 1900s, the work of English neurologist Henry Head (1861–1940) further reinforced the specificity theory. Head was frustrated by the unreliability of his patient-subjects relating exact details of their experience of cut and recovering nerves. In 1903 he enlisted the help of surgeon William Rivers in dividing the superficial ramus of the radial nerve in his own arm. Until 1907, Head and Rivers charted the loss of sensation and its slow recovery in Head's arm. Head noted two clear stages in recovery. First, he noticed a crude and generalized return of feeling, which he called the protopathic stage. The feelings were not well-localized nor could Head determine levels of intensity; he could distinguish hot and cold, touch and pressure, and whether a sensation was pleasant or unpleasant, but without sufficient detail to tell, for instance, what kind of object was touching his arm. The second stage, the return of normal cutaneous sensations with their usual levels of discrimination, was slower. Head called this 'epicritic sensitivity'.

which sense had been lost; the other was to apply different types of stimulus to areas of skin, then examine the nerve endings found there and compare them with the sensation reported by the subject. Finding that there are spots specific to discrete types of stimulus was easier than matching receptors to functions.

The Swedish physiologist Magnus Blix experimented on his own skin in the early 1880s, applying a very small electric current to tiny areas one at a time and noticing the kind of sensation it produced. He reported that the same stimulus could, in different places, produce feelings of touch, heat or cold. He thought it more likely that he was discovering the specificity of nerves themselves (see page 131) rather than the specificity of the receptors. He developed further experiments which revealed that heat and cold are detected at different spots on the skin (so by different nerves

or receptors) and there is also a distinction between the perception of pressure and pain.

Alfred Goldscheider confirmed Blix's findings with his own experiments in 1884, but also showed that sometimes (with difficulty) he could produce the same sensation of temperature by stimulating a nerve fibre under the skin rather than its receptive end. He distinguished three different intensities of touch – tickle (the lightest), touch and pain. He did not believe pain needed its own receptor, but found that some (though not all) touch receptors would register pain if the pressure of the touch were increased. He theorized that pain occurred when very strong stimulations caused neural discharge to overflow into a special pathway in the grey substance of the spinal cord. The receptors for painful stimuli – nociceptors – are now considered distinct from those that deal with touch and vibration and with heat and cold. The very specific case of pain is dealt with in the next chapter.

The American Henry Donaldson also duplicated Blix's experiment. He then had a surgeon remove pieces of his own skin in areas where he had identified spots sensitive to heat and cold, but was unable to identify receptors.

Nerve specificity

The idea of specificity was not restricted to the receptors, but extended to the nerves. Weber first tried to block nerves carrying signals relating to cold in 1847 and other researchers followed, trying to block different types of sensation. This was based on the belief that there are nerves separately dedicated to feelings such as heat, cold and touch – known as nerve specificity. Alexandre Herzen showed in 1885 that if a tourniquet is used to block nerves, sensations are lost in the sequence cold, touch, heat, superficial pain and then deep pain. When the tourniquet was removed, the sensations returned in the reverse sequence. This makes sense in terms of protecting the organism – pain is the most important sensation to preserve as it prompts us to move away from harm.

Attention to specific fibres and research into them began in the 20th century. In 1916, Stephen Ranson reported that cutting certain thin nerve fibres in a cat eliminated the sensation of pain. In 1929, more comprehensive work was carried out in the USA by Joseph Erlanger and Herbert Gasser, who found that nerve fibres fall into three groups, with thickness correlated to speed of transmission. The thickest are A fibres, with the fastest transmission speed, and the thinnest and slowest are C fibres. Pain is associated with the smallest fibres, hot and cold with those of intermediate size, and touch, muscle, sense and movement with the largest fibres. Although the fibres differ, the nature of the signal transmitted is the same.

Inside and out

Although we have focused here on the sensory receptors in the skin, the full range of somatosensory function includes receptors in the joints, bones, muscle, and other internal parts and organs which also keep us aware of our position, and which sense vibration and movement. In addition, there are receptors

and signals of which we are never or rarely aware, typically forming part of the parasympathetic nervous system and involved in keeping our body functioning (breathing, pumping blood, digesting food, and so on).

From sense to sensorium

While it's obvious that sensory organs are specialized to respond to different stimuli, it's not immediately apparent whether the method of communication with the brain is specialized. Is the communication between the tongue and the brain the same as that between the eye and the brain? Does the information which produces sense-impressions of different types differ in its method of perception, its method of transmission, or only in its interpretation within the brain?

Energy of different flavours

The Scots neurologist and surgeon Charles Bell wrote in 1811 that 'each organ of sense is provided with a capacity for receiving certain changes to be played upon it, as it were, yet each is utterly incapable of receiving the impression intended for another organ of sensation.' This is perhaps the first suggestion that it is not the stimulus that determines our sensory experience but the organ that is stimulated. Bell's work was not well known, however.

In 1826, the German physiologist Johannes Müller outlined his theory of 'specific nerve energies' and made the same point as Bell, but related it not just to the sensory organs but to the nerves serving them. He stated that no matter how a sensory organ is stimulated, it can only pass on information to the brain of the type that it is routinely used for. For example, as Isaac Newton noticed (see page 118–19) the eye responds both to light and to pressure by producing visual sensations.

Müller suggested that the messages carried by the nerves from different sensory organs have a quality specific to the organ. Müller was a vitalist, and believed that living organisms are in possession of some type of life-energy which cannot be fully accounted for by science. Even so, he laid the foundations for the modern, integrated approach to physiology that brings human and comparative anatomy, chemistry and aspects of physics to bear on physiology. He mentored and influenced many of the great physiologists of the 19th century, including du Bois-Reymond, Helmholtz and Schwann.

Although he was wrong about the principle of different types of energy, Müller took an important step in the right

SYNAESTHESIA

Synaesthesia is a naturally occurring example of what can happen when a single stimulus is processed by more than one sensory area of the brain. People with synaesthesia may, for example, see colours and shapes when they hear a sound. Synaesthesia can take many forms, and can be present from birth or develop after brain injury. The mechanism and causes are not properly understood. Many of those who experience this consider it a gift rather than a disability.

direction by surmising that it is not the nature of the external stimulus (light, sound, and so on) that determines the sensory impression created in the brain. Helmholtz developed Müller's doctrine in the 1850s and 1860s, suggesting that specific nerve energies might account for the different perception of different colours, tones, and so on. This led him to the suggestion of 5,000 different types of receptor for sounds.

Same signal, different experience

The opposite view was that the type of transmission is exactly the same in all types of nerve fibre. Du Bois-Reymond – one of Müller's students – preferred this view. It led him to suggest that if we could swap the nerves from eyes and ears to the brain, so that the auditory nerve went to the visual cortex and vice versa, we would be able to see thunder with our ears and hear the lightning flash with our eyes.

In 1912, Lord Adrian demonstrated that the type of energy carried by all the nerves is indeed of exactly the same type: electrical energy in the form of action potentials. The way in which we experience a stimulus depends on the part of the brain to which the stimulus is delivered by the nerves. So information coming from the optic nerve will always be interpreted in a visual form.

Johannes Müller

Putting it all together

The early model of how the sensorium builds experience from sensory input suggested

> *'The same cause, such as electricity, can simultaneously affect all sensory organs, since they are all sensitive to it; and yet, every sensory nerve reacts to it differently; one nerve perceives it as light, another hears its sound, another one smells it; another tastes the electricity, and another one feels it as pain and shock. One nerve perceives a luminous picture through mechanical irritation, another one hears it as buzzing, another one senses it as pain. . . . He who feels compelled to consider the consequences of these facts cannot but realize that the specific sensibility of nerves for certain impressions is not enough, since all nerves are sensitive to the same cause but react to the same cause in different ways.'*
>
> Johannes Müller, 1835

that all the sensory nerves come into the 'common sense' area of the brain in the first cell (see page 18). Here the brain supposedly put together the different aspects of how something is perceived – say, the sight, sound and feel of a dog – to make a composite image or experience of 'dogness'. But this was, of course, entirely hypothetical.

Perception and apperception

Gottfried Leibniz broke down whole perceptions into many infinitely tiny ones, which he called 'petites perceptions'. (Leibniz was, with Newton, an originator of differential calculus, which depends on breaking larger phenomena into minute portions.) He gave the example of the sound we hear when listening to waves crashing on the shore. The sound comes from lots of tiny movements of small bodies of water; added together, the noise is loud, but if we were to single out just one moving drop it would be too quiet to hear.

Even though these little movements would seem to be nothing, added together we are aware of the sea, and a sensation can't be made up from a lot of nothings. Leibniz proposed that there is a tipping point, or threshold, which he called the *limen*, above which we are aware of a phenomenon and below which we are oblivious. When the mass of micro-perceptions is sufficient to be noticeable, he called it apperception – so apperception is the point of awareness. Below the threshold, we remain unconscious of the micro-perceptions; this was possibly the first suggestion of an unconscious mind. It was this concept of the threshold of perception that Weber explored and labelled the 'just noticeable difference'.

What you see is not what you get

Interestingly, when Müller proposed his doctrine of specific energies, he made the point that what the brain receives is not information about the outside world, but information about the state of the nerves. It is this that is interpreted to give the perception of sound, light, pressure or whatever that particular nerve is capable of conveying: 'Sensation consists in the sensorium receiving . . . a knowledge of certain qualities or conditions not of external bodies but of the nerves of sense themselves.'

As we have seen, information from the sensory nerves goes to different

The brain puts together different types of sensory input to construct the sense of 'dogness'.

FEELING AND FEELINGS

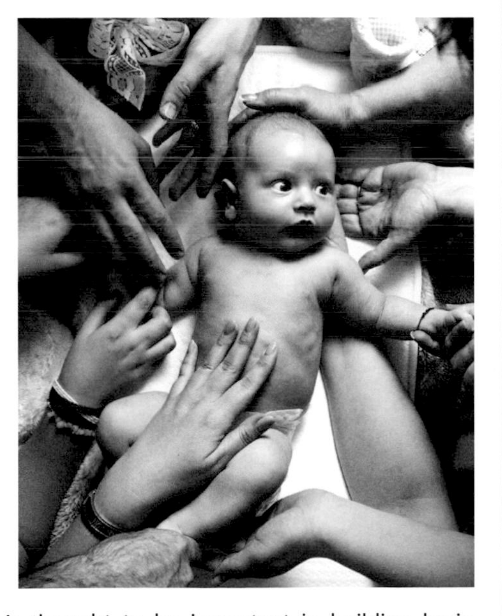

Not all nerve fibres are alike. Myelinated nerves carry an action potential more quickly than unmyelinated nerves, and larger diameter nerves carry a signal more quickly than thinner nerves. Touch perception is transmitted to the brain by two types of fibre, one fast and one slow, which go to different parts of the brain. The first signal helps localize the touch, and goes to the somatosensory cortex. The second goes to the insular cortex, which deals with emotion and prompts the emotional response to the touch without engaging conscious thought. This mechanism is thought to be important in building brain connections in newborn infants, who fail to thrive if deprived of a gentle touch.

THE 'RUBBER HAND' ILLUSION

Our senses make us aware of where each part of the body is and what it is in contact with – but they can be fooled. The rubber hand illusion demonstrates this. A subject's hand is hidden from them behind a screen and a rubber hand placed in such a position that it looks like their own. Their real hand and the rubber hand are stroked at the same time and in the same way with a brush. After a short time, the person comes to feel that the rubber hand is their own – to the extent that they flinch if the rubber hand is suddenly struck with a hammer. The rubber hand illusion was first described in 1998. It seems to show the brain's plasticity allowing it to reconfigure the image of the body and its extent.

locations in the cortex that were identified individually during the course of the 19th and 20th centuries. In 1945, American neurobiologist Roger Sperry showed it is the location in the brain where the nerves carry the information that determines how we experience a stimulus. He experimented on animals, severing nerves and rerouting them; in every case, the animals acted in accordance with the area of the brain stimulated. For instance, a rat with the nerves for the left and right legs switched in the brain would always lift the left leg if the right leg was stimulated with an electric shock. Sperry found that no matter how much time he allowed for animals to recover, they never readjusted. He concluded that some aspects of the brain's control are hard-wired and plasticity does not step in to fix things.

The conclusion is that while the sensory receptors are specialized to respond to different types of stimulus, the means of transmitting information from them is identical – there is no specificity of nerve energy. The way that a stimulus is interpreted by the brain depends only on where the nerve carrying the information ends. If the nerve ends in the visual cortex, the information will be perceived as visual, even if it was produced by pressure on the eyeball rather than light on the retina, for instance, or even if the nerve has been rerouted from the ear. We are perceiving, as Müller recognized nearly 200 years ago, the state of our own nerves rather than the outside world.

The whole picture

Our senses are our means of gathering information from the world outside the body – or outside the mind, since some are also involved in communicating within the body. They provide an interface not just between the mind and the physical environment (internal and external); they are often also the meeting point of mental activity and physical activity. They can bring pleasure and pain, in both physical and emotional domains. To achieve this, the brain creates and maintains a representation of the body.

Making sense of the senses

Our senses allow us to experience a lot of things, from watching a movie or listening to music to relishing good food or responding to a hug. And they provide information that can be acted on to keep the body safe or sustain it. Seeing or hearing a predator can trigger the urgent desire to hide or flee; tasting a fruit as bitter can trigger the urge to spit it out in case it is toxic. The response to sensory input might be a willed or involuntary physical action. Or it might be an involuntary internal response: producing saliva and gastric juices in response to smelling or tasting food, for instance. A single type of sensory input can have more than one effect if it goes to more than one part of the brain. One type of sensory input which we have not dealt with here is pain. How we perceive pain is more complex than the other senses and will be discussed separately, in the next chapter.

Taste, touch, smell, sight, hearing – our perception of the world largely depends upon our senses.

CHAPTER 7

A bit of a PAIN

'Nothing can properly be called pain unless it is consciously perceived as such.'

William Livingston, 1943

In the past, physical pain was thought to be an exaggerated response to touch, or an overloading of the sense organs in the skin, but it is quite distinct from the other sensations. The perception of pain varies more between individuals than the other senses; it also varies in the same individual at different times and under different circumstances. This is because the brain plays a greater part in constructing, rather than in simply interpreting, the experience.

The ability of a fakir to endure experiences that most people would find painful underlines the subjective nature of pain.

Protective pain

From a neuroscience perspective, pain is extremely interesting. Throughout history, people have sought ways of avoiding or alleviating pain, yet it is a vital response. People who feel no physical pain can sustain terrible, even life-threatening, injuries or illness without recognizing the danger they are in – pain is a useful safety mechanism.

Feeling pain

The Ancient Greeks did not assign the brain any part in the production of pain. Plato and Aristotle considered it to be an emotion, a 'passion of the soul', rather than a physical sensation. Aristotle believed that pain and pleasure moved through the body in the blood to reach the heart. Consequently, he thought that the places best supplied with blood were also the most sensitive.

From the 5th century BC, Hippocrates accounted for pain as a manifestation of the imbalance of humours in the body. Galen accepted this, but went on to say that three things were needed for a person to feel pain: an organ that receives an impression of pain, a passageway connecting the organ to the brain to carry the impression, and an organizational centre in the brain which recognizes the pain. He believed the brain to be the most important organ involved in the sensation of pain and recognized four categories of physical pain: pulsating (throbbing), lancinating (stabbing), weighty (heavy or dull) and stretching (cramping or taut).

In the 11th century, Ibn Sina extended Galen's categories of pain to 15, many of which (like Galen's original four) correlate with categories recognized in the pain questionnaire used by doctors today to rate pain. Ibn Sina noted that pain does not

DEFINING PAIN

We use the same word, 'pain', for localized physical hurt caused by injury, for the generalized discomfort of disease and for emotional distress. Most neurological discussion of the origins of pain focuses on the first two, and particularly the pain of injury.

Pain prevents the player risking further damage by continuing with the game after injury.

require the presence of an ongoing injury – it can continue after the original stimulus causing the pain has been removed.

Causes and mechanics of pain

From ancient times until at least the 17th century, pain was more likely to be seen as divinely inflicted, either as punishment or trial, than as something that could be understood rationally. So the pain of disease was not to be explained in terms of exactly how and why things had gone wrong with the body, but was often seen as visited on individuals by a vengeful deity or, in some cultures, as the result of a curse or bewitchment. Pain could even be seen as a form of penance that might speed the unfortunate sufferer's passage through purgatory – by getting some of the suffering done in advance. This was no doubt a useful line for physicians and priests alike to peddle to people they could neither help nor console: it might be bad now, but if it could save a few centuries of torment it was worthwhile, so the patient should not grumble too much and unrealistically crave respite.

Pain in body and brain

From the time of Epicurus (342–270BC), it was generally believed that the severity of pain correlates with the extent of an injury. This notion survived largely intact until

In banishing Adam and Eve from Paradise, the God of Genesis *cursed Eve (and thereafter all women) with pain during childbirth. Pain has long been considered a punishment or gift from God which has sometimes translated into a reason for refusing or withholding pain relief during labour.*

ATOMS OF PAIN

The Greek philosopher Democritus taught that all matter comprises very tiny particles, called atoms though not quite the same as the ones physicists describe today. He accounted for sensations by having objects shed some of their atoms, which then make their way into the body. In the case of pain, sharp or hooked atoms entangle with the atoms of the soul.

the 1960s. Yet it ignores several factors, not least of which is the importance of the psychological dimension in the experience of pain. Of all the senses, pain is the one that is most subjective: each person's experience of a similar level of injury can be very different. Furthermore, an individual may experience a minor injury – a paper cut or mouth ulcer, for example – as more painful than a severe wound.

The theory of pain explained by Galen's model of a receptor, a pathway and a perception centre in the brain, doesn't allow for types of pain that don't correlate directly to an injury. For example, the phantom limb pain often described by amputees and the all-too-common experience of chronic pain for which no systemic cause can be found fall into this category. Even today, some people consider these not to be 'real' pain. It remains a matter of debate whether 'real' pain must have an identifiable physical cause or whether it can also be a subjective mental experience of suffering.

The transmission of nerve signals associated with 'noxious' (damaging or unpleasant stimuli) has its own name – nociception. In terms of neurology, it is easier to begin with the kind of pain that is produced in response to the stimulation of nociceptors.

Working with pain

The first constructive neurological work on pain was carried out by Johannes Müller. He believed that there are specific nerve fibres for pain and special receptors for picking up painful sensations. He maintained that pain could only be felt as a result of stimulation of those sensory nerves that transmit pain signals. This fitted pain into the larger nerve-specificity model.

The opposing view – that there are no special nerves or receptors for pain – had a long history. Aristotle was among those who argued that pain is the result of extreme stimulation of any type. It can come from excess heat, noise, bright light or many other extreme effects and can be transmitted to the soul in many ways. This model endured in principle for more than 2,000 years, varying only in its details.

In the 17th century, Descartes was the first to begin seeing pain as something internal with rational explanations both for its origins and transmission. In his mechanistic view of the human body, pain was the result of the machinery being unbalanced or out of kilter in some way. He distinguished between the mechanism of pain – a stimulus that causes the body to withdraw from something harmful – and the mental experience of pain. In 1644, he described pain as a 'disturbance' that travelled from the periphery along the nerves to the brain.

In 1874, Wilhelm Erb stated that any type of sensory receptor can produce a pain signal if stimulated sufficiently strongly. Yet in 1858 Moritz Schiff showed that different pathways along the spinal cord are associated with pain and touch, supporting the idea of specificity. The two models – intensity of stimulus and specificity – coexisted for a while, but with intensity favoured more by psychologists than by neurologists. By the end of the 19th century, most experts accepted the specificity theory. This was further reinforced by Henry Head's experiment on the nerves of his arm in the first decade of the 20th century (see page 130).

Neither Blix nor Goldscheider (see pages 130–31) included pain in their investigation of receptors in the skin in the 1880s. Ten years after their work, Max von Frey proposed that pain is a separate modality and is associated with free nerve endings. Von Frey's ideas were popular, probably because they were quite simple – one type of receptor for each type of cutaneous sense – but they were wrong in their detail. Over subsequent years, other researchers found additional types of receptor. They also observed that areas with free nerve endings could be cut without causing pain. In short, the whole situation was a lot more complicated than it looked.

During the late 19th and early 20th centuries, physiologists tried to trace the nerves that carry cutaneous feelings, including pain, and identify the areas of the brain responsible for turning nerve transmissions into experience.

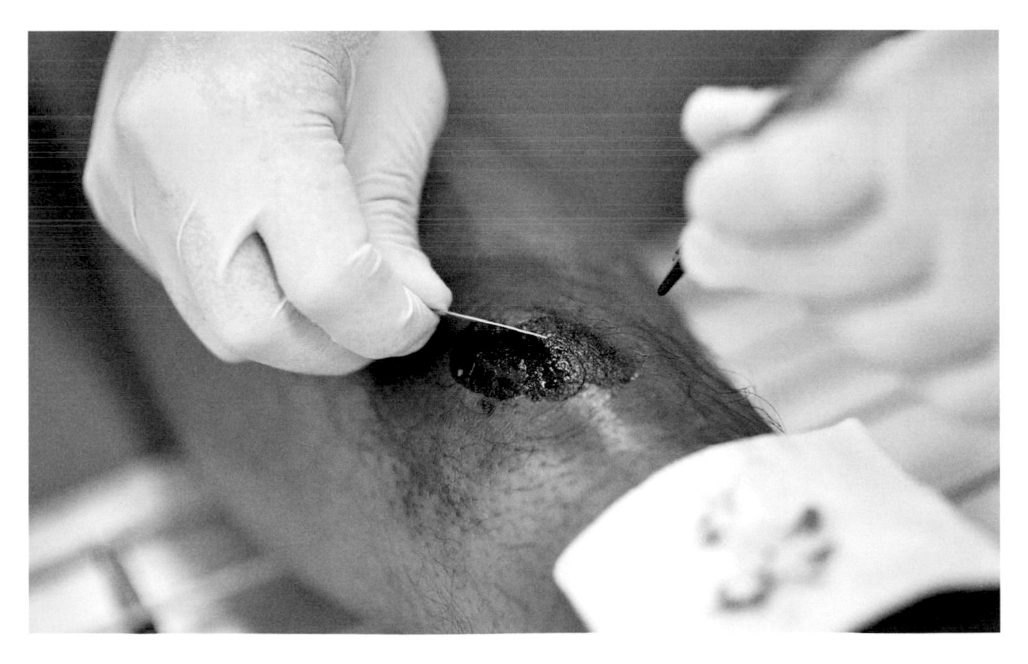

Burns are extremely painful, which must have seemed to support the idea that pain could be produced by extreme stimulation of other types of receptor.

TYPES OF PAIN

Nociceptors – specialized receptors for intense stimuli – were first described by Charles Sherrington in 1906. Different types of nociceptor respond to different sorts of noxious stimuli: high temperature, extreme cold, intense pressure or toxic chemicals. Since the work of the English neurophysiologist Thomas Lewis in 1942, nociceptors have been associated with two different types of neuron. These were identified by the Canadian psychologist Ronald Melzack: one type is myelinated and transmits signals rapidly (A fibres); the other is unmyelinated and carries signals more slowly (C fibres). This leads to two different types of pain experience – first a sharp, sudden pain, and then a duller but enduring pain. (Think of the stabbing pain of a cut, followed by the ache that sets in afterwards.) The first type is associated with acute pain and the second with chronic pain.

All or nothing

As with other neurons, those nerve fibres carrying a pain signal are either fired or not – it's an all-or-nothing system. This works in sensory neurons responding to a stimulus and in motor neurons connecting to muscle tissue and causing muscles to contract. The principle was first noted by the American physiologist Henry Pickering Bowditch in 1871, when working on the contraction of heart muscles. As it is a binary response, there is no variation in the strength of a signal – either a neuron fires (initiates an action potential) or it does not. But if there is insufficient stimulus to produce a response immediately, there is still some effect on the balance of ions inside and outside the neuron and, if the stimulus continues, this effect can build up to a tipping point and the neuron fires.

The intensity of what we experience is not related to the intensity of the stimulus

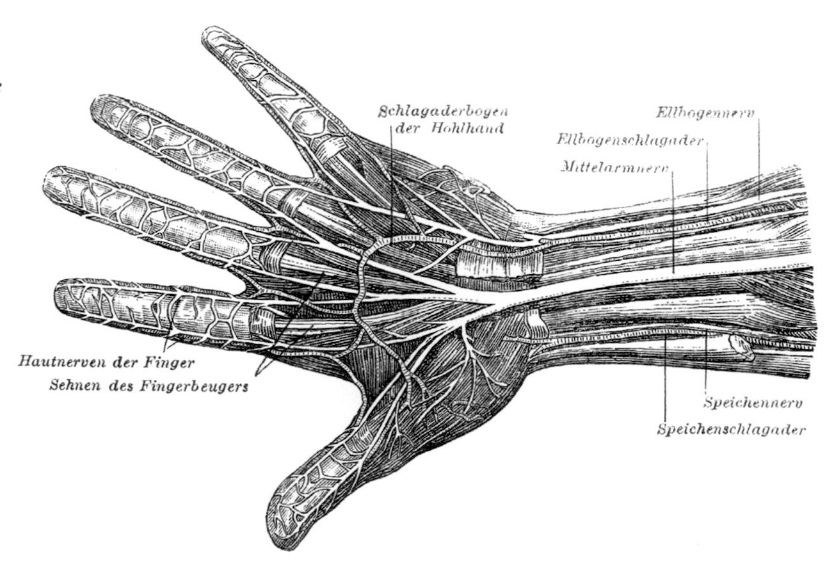

A 19th-century illustration of the innervation of the hand.

of individual neurons, but to the number of neurons stimulated. If we look at a bright light, many neurons in the retina fire, but if we look at a dim light, fewer are affected; this is because fewer photons are falling on the retina so fewer neurons will reach the stimulus threshold at which they fire. The same is true of a pain trigger, which is why there may be a mismatch between the severity of an injury and the experience of pain. If you drive a thin needle deep into your leg it will hurt less than if you make a long but shallow cut, simply because fewer pain-sensing nerves fall in the path of the needle.

The threshold effect is not limited to the firing of individual neurons. We would be overwhelmed if every time a single neuron was triggered it produced an experience of pain. Instead, there must be a sufficient number of neurons firing at once to cause a signal to be sent to the brain to register a feeling of pain (see page 146). This was discovered in principle by the German pathologist Bernhard Naunyn in 1889. He applied very rapid but small stimuli to patients, below their threshold for noticing touch, and found that over a short period an unnoticeable stimulus led to unbearable pain. He was stimulating nerves several hundred times a second over a period of 6–20 seconds. He concluded that pain is summative: if enough stimulus occurs over time, a pain response will be triggered.

Specificity and intensity combined

In 1943, American physiologist William Livingston proposed a summation theory that built on the findings of Naunyn and others. He proposed that as signals produced by a painful stimulus reach

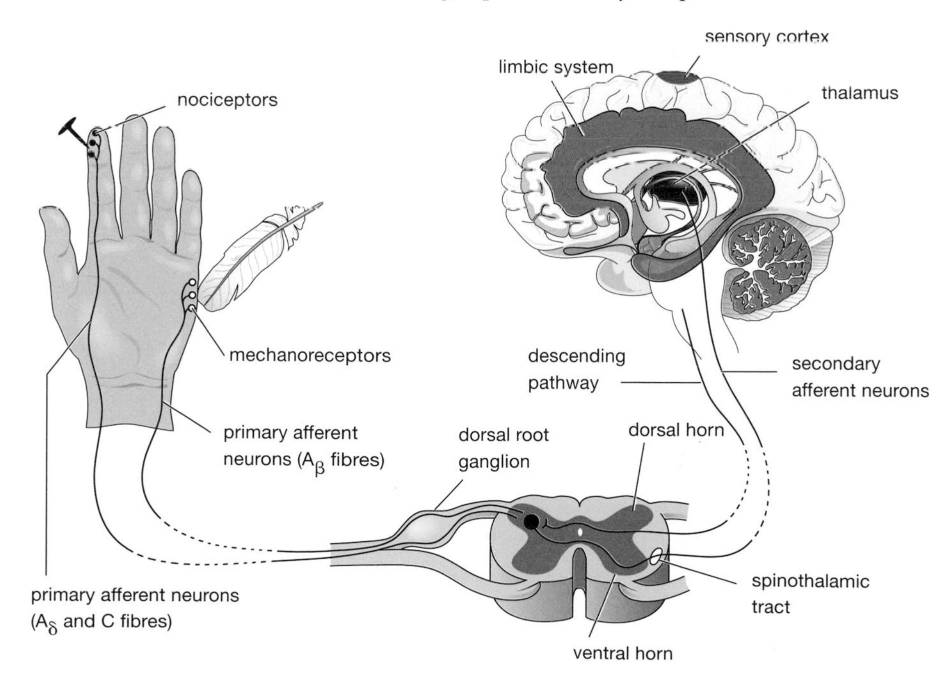

the spinal cord, they build up in a loop of activity in interneurons until a threshold is reached. At that point, a signal to the brain is triggered and pain registers. The interneuron activity, he suggested, also spreads to other spinal nerves and can trigger more activity including motor and sympathetic system responses, together with fear and other emotional responses.

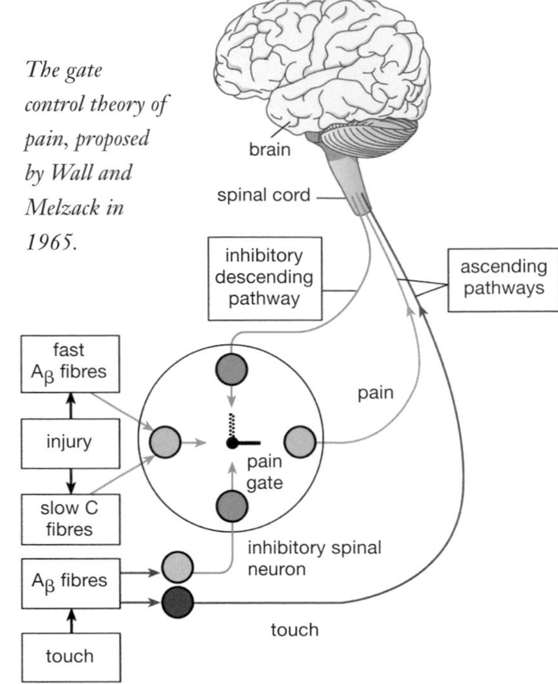

The gate control theory of pain, proposed by Wall and Melzack in 1965.

Rub it better

The Dutch researcher Willem Noordenbos (1910–90) noticed in 1953 that signals carried by large nerve fibres can effectively silence those carried from the same area by thinner fibres. The intensity of pain we feel depends on the degree of stimulation of thin and thicker fibres, carrying pain and touch/pressure signals. One result of this is that pain is genuinely reduced by rubbing the site of an injury, the transmission of touch/pressure signals damping down the effect of the pain signals.

Letting pain through

In 1965, British neuroscientist Patrick Wall (1925–2001) and Canadian psychologist Ronald Melzack (b.1929) gave a more detailed explanation of this effect with their theory that pain is 'gated'. This maintains that signals have to pass 'gates' to get from the spinal cord to the brain. Signals from nociceptors are either blocked or allowed through these 'gates', determining whether we feel pain and/or how much pain we feel. Melzack's work was based on the findings of Lord Adrian and Charles Sherrington using a galvanometer to measure the action potential of different nerve fibres, and on Gasser and Erlanger's identification of the three types of nerve fibre, A, C and B. Type B fibres are partly myelinated, and are intermediary between types A and C.

The gating system works by impeding (or not) the action of inhibitory interneurons. If that sounds a bit like a double negative, it is. The inhibitory interneurons block the passage of a signal up the spinal cord to the brain by preventing the release of a neurotransmitter used to pass on signals from the pain neurons, so when they are working a signal does not get through. When their action is impeded, they can no longer stop the signal, so it passes through. Stimulating the thin C-fibre neurons that carry pain signals impedes the inhibitory neurons, so the pain signal can be sent to the brain. But stimulating thick A-fibres promotes the action of the

inhibitory neurons. This means that if there is more action from the thick fibres, the pain signal will be inhibited or reduced. It is for this reason that rubbing an injury or applying heat or cold to an injury tends to reduce the pain. It's also the basis on which the TENS (transcutaneous electrical nerve stimulation) machine works to relieve pain.

Finally, the brain sends top-down messages to determine whether pain signals are allowed through. Melzack proposed another route for pain signals which allows them to short-circuit the gating mechanism and go straight to the brain. The brain would then make a decision about whether to block (or allow) the inhibitory action, and could send a signal to the spinal cord to do this. This top-down control of pain signals may explain the well attested impact of psychology on pain. In an emergency, people sometimes don't experience pain, even from severe injuries. This is because the body has more important things to get done – escaping from a dangerous situation, for example. People who are busy or relaxed often experience pain to a lesser degree than those who are idle or stressed.

INNER PAIN

Sensory and pain receptors are not only found on the surface of the body, such as the skin, but also at many points inside the body, including joints, bones and some internal organs. Other organs have no receptors that can signal damage, which can lead to disease becoming dangerously advanced before the sufferer is aware of it.

Pain and the mind

Explanations of the physiology of pain can give a good account of how the body senses and transmits potentially dangerous stimuli, but it can all be confounded by the mind. It is the brain that assembles the experience of pain, and in doing so it draws on past experience, expectations and a host of other complicating factors. As the most subjective of the cutaneous senses, pain is a minefield for researchers, especially as measuring pain relies exclusively on self-reporting. It is not possible for someone to literally 'feel'

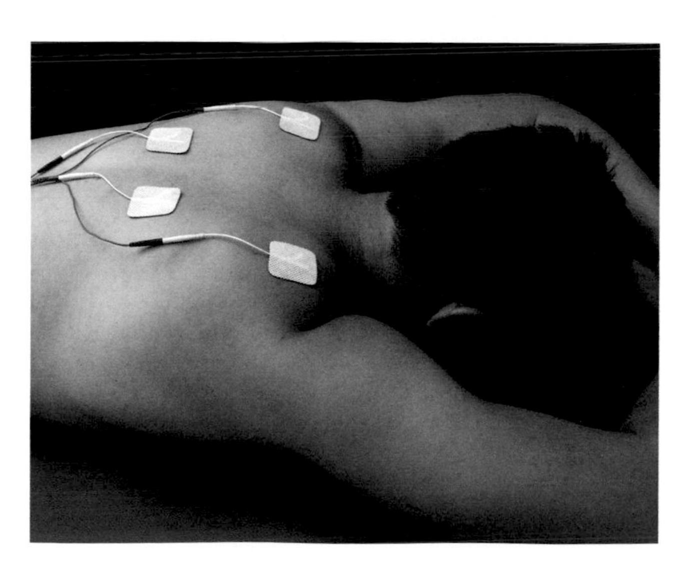

A TENS machine is often used to alleviate muscle pain.

REFERRED PAIN

Some 'shared' nerve pathways carry pain signals from different parts of the body. In this situation the brain has no way of recognizing the origin of a painful signal so pain may be experienced in a different part of the body from the one affected. A common example is heart attack, which may be experienced as pain in the jaw or arm, even though it is damage to the heart muscle alone that has triggered the pain sensation. In this case the brain has attributed the signal as coming from areas where signals most commonly arise (jaw and arm) rather than the heart, which rarely gives rise to such signals.

another's pain. Pain can be felt in the 'wrong' place – that is, not the place that is damaged or diseased – or even in a place that no longer exists.

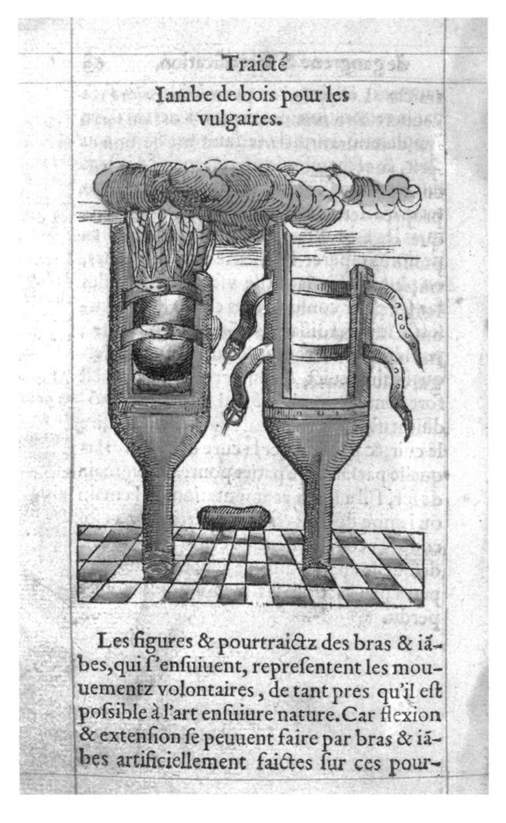
Traicté

Iambe de bois pour les vulgaires.

Les figures & pourtraictz des bras & iã-bes,qui ſ'enſuiuent, repreſentent les mou-uementz volontaires, de tant pres qu'il eſt poſsible à l'art enſuiure nature.Car flexion & extenſion ſe peuuent faire par bras & iã-bes artificiellement faictes ſur ces pour-

Wooden artificial legs used by Ambroise Paré to help amputee patients.

Ghostly pain

One of the more fascinating facets of pain, and one that has proved particularly fruitful for researchers, is the phantom limb pain experienced by people who have lost an arm or a leg. This was first reported by the French surgeon Ambroise Paré in 1551 and mentioned also by Descartes. Around 60–80 per cent of people who have lost a limb say they continue to feel sensations emanating from it, most commonly pain.

> *'[Pain is] an unpleasant sensory and emotional experience associated with actual or potential tissue damage, or described in terms of such damage.'*
>
> International Association for the Study of Pain, 1975

The American surgeon Silas Mitchell, who worked at 'Stump Hospital' during the American Civil War of 1861–5, reported that 86 of the 90 amputees he examined complained of phantom limb pain. With this large sample of subjects he was able to gain a general picture: the soldiers reported a phantom limb that was usually shorter than

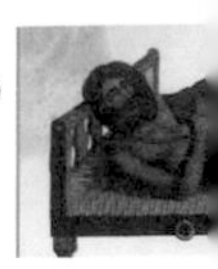

This lucky patient of pioneering saint-surgeons Cosmas and Damian will be spared phantom limb pain by their miraculous transplant operation.

the original limb and not always complete. They complained of considerable pain, which could be brought on by even minor stimuli such as the wind blowing over them. Wearing a prosthetic limb could also trigger pain, which proved resistant to treatment such as cauterizing the nerves, acupuncture or drugs. Some soldiers even had more of the limb removed in the hope of respite; they were usually disappointed.

Mitchell and others believed the pain was caused by irritation of the nerves that had been severed during amputation. This suggested that the stimulation of these cut nerves sent messages to the brain, which were interpreted as emanating from the original nerve endings, even though these were now missing. Treatments were brutal and rarely successful. An alternative to shortening the stump was to cut the sensory nerves between the stump and spinal cord, or even remove the part of the thalamus that was the ultimate destination of the signals.

The 'irritated nerves' theory was finally shown to be wrong in the 1980s by Melzack.

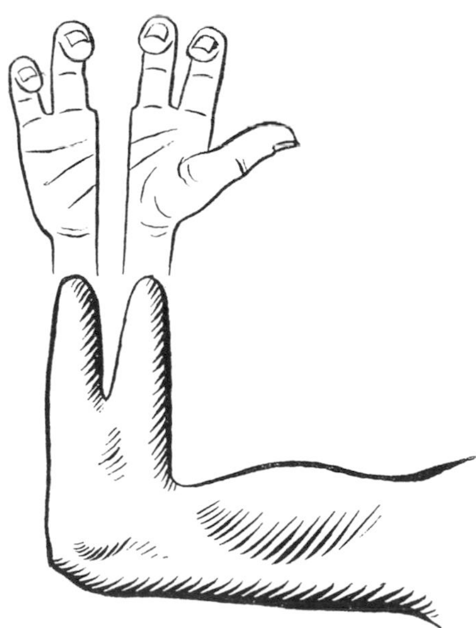

Melzack noted that people who had the stump of an arm split to enable them to grasp objects found that their phantom hand felt correspondingly split.

He began by tackling the commonsense model of pain that had underlain all previous research since the time of Descartes – the idea that pain is felt in the injured part and the pain signal is transmitted to the brain where it is turned into the experience of pain. This all seems entirely logical until it is confronted with phantom limb pain and other types of chronic pain which are not directly correlated with physical damage. Melzack also pointed to the experience of people paralyzed by the severance of nerves in the spine who nevertheless continued to experience phantom limbs. They could not be experiencing irritation of nerve endings in the 'phantom' limb as it was physically still present, but had no neurological connection with the brain.

IT HURTS LESS IF YOU WATCH

Research in the early 21st century has found that a potentially painful procedure hurts less if we can see it – which is a good reason for watching when you are given an injection or have blood taken. In 2008, fMRI scanning showed that areas of the brain involved in processing pain are also involved in processing the size of visual inputs.

Ingredients of pain

Melzack described a neuromatrix model of pain which takes the complex task of constructing pain away from the peripheral nervous system and allocates it to the central nervous system. Pain is not, then, produced by tissue damage, but by various parts of the CNS acting together on information from the PNS and the environment. According to his theory, and subsequent research, the parts of the CNS involved are:

- spinal cord
- brain stem and thalamus
- parts of the limbic system, including the hypothalamus, amygdala, hippocampus and anterior cingulate cortex
- insular cortex
- somatosensory cortex
- motor cortex
- prefrontal cortex

Between them, they produce the sensory, emotional, cognitive, behavioural, motor and conscious aspects of the experience of pain. The model explains why pain does not always correlate to the amount or severity of tissue damage – or even to there being any damage (or tissue, in the case of phantom limb pain). It also explains how the placebo effect works (see page 152) and how people sometimes fail to notice or feel traumatic injuries in times of stress, such as on the battlefield or in other dangerous situations.

Melzack maintains that the neuromatrix is, at least in part, genetically determined. Children born without a limb can have a phantom limb. This suggests that the neuromatrix combines elements of genetic make-up and learning. The importance of learning is clear from the experience of people with phantom limbs who continue to feel, for instance, the presence of a painful bunion or a tight ring.

When chronic pain is unrelated to obvious tissue damage, modern pain management strategies often combine exercise, cognitive behavioural therapy and

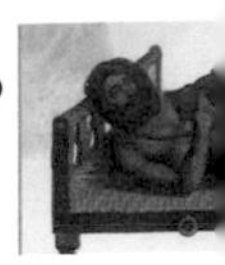

REWRITING THE BODY MAP

The Indian neuroscientist Vilayanur Ramachandran (b.1951) suggested that after an amputation, the map of the body held in the somatosensory cortex is rewritten (see page 148). He says that the 'rewiring' of the somatosensory cortex explains why for some amputees being stroked on the face is experienced as a feeling in the phantom limb.

In 1994, Ramachandran pioneered a type of therapy for phantom limb pain, called mirror-box therapy. The patient inserts their remaining limb into a box with a mirror which creates a reflection representing the missing limb. By moving the real limb, some patients are able to relieve pain in the phantom limb. (Phantom limbs are often 'held' and stuck in awkward, painful positions.) Mirror therapy does not help all people suffering from phantom limb pain, and research into when it is effective continues.

medication. These rely on the neuromatrix model and the idea that the body is constructing pain in response to a number of stimuli and internal patterns.

Mostly in the mind?

Brain imaging carried out when a patient is enduring painful experiences reveals that 300–400 areas of the brain are involved in sensing or constructing pain – it's a very complex system. Studies in 2004 found that brain activity related to pain was reduced if a person was distracted or meditated during a painful stimulus, confirming findings from observation and common experience.

A study published in 2012 revealed how patients' beliefs and expectations can be borne out by their experience of pain, independent of what is actually going on in their bodies. Volunteers received a burning sensation to the arm while hooked up to an intravenous line that they were told would deliver a painkiller. They were notified when the painkiller was started, and when it was later stopped. They were told they might feel worse pain when analgesia was removed. The average ratings for pain were: first burned – 6; soon after – 5; with analgesia – 2; after removal of analgesia – 6. In fact, analgesia was started at the 'soon after' stage, showing only a very small

impact on the experience of pain when volunteers didn't know they were receiving it. It continued after they were told it had been stopped, yet had no effect when they didn't believe they were receiving it. This is related to the well-known 'placebo' effect.

Placebo, nocebo

Medicines which look like actual medicines but have no pharmacologically active ingredients are called placebos. They are usually given during clinical trials to test the effectiveness of a newly developed drug. One group of patients receives the active drug and another – control – group receives the placebo. The outcomes of the two groups are then compared. In theory, those patients receiving the placebo should show no improvement in their condition. Yet, in practice, patients receiving the placebo may often enjoy improvement or recovery rates comparable with patients receiving pharmacologically active medicines. The placebo effect can be very powerful, as the study of analgesia showed. The opposite of the placebo effect is called the nocebo effect, when patients experience a noxious effect from a pill that contains no active ingredient (see box below).

It doesn't even seem to matter if the patient finds out that the medicine which works for them is a placebo. A study in 2015 showed that if patients took a placebo for four days, believing it to be a painkiller, it continued to work for them, reducing their pain even after they had been convincingly told it was a placebo. The effect didn't work in patients who received the placebo for one day only, suggesting that an element of conditioning is involved.

Something equivalent to the nocebo effect is thought to lie behind confirmed cases of people responding to curses by falling ill and dying. In the vodun (or voodoo) tradition of Haiti (and parts of Africa), people who have been cursed may die unless the 'curse' is lifted by a powerful magician or witchdoctor. Conventional medicine cannot usually save them because there is no systemic disease to treat. This effect goes far beyond feeling pain with no physical correlate: the patient's belief causes complex physical effects that are generally considered to be beyond conscious control.

DEADLY SUGAR PILLS

In 2007, a man suffering from clinical depression agreed to take part in trials of an antidepressant. He was unaware that he had been assigned to a control group and was taking dummy capsules with no antidepressant in them. Feeling suicidal, the man took an overdose of 29 capsules. His blood pressure dropped to a dangerously low level and he needed intravenous fluids to keep him alive. When he was told that he was receiving the placebo, his physical condition rapidly returned to normal.

Brain and pain

It appears that of the three entities Galen thought were necessary to produce pain, only one is really required; it is the organizational centre in the brain which decides whether or not we will express pain.

BEARING UNBEARABLE PAIN

Some societies insist on adolescents undergoing painful initiation ceremonies to mark their transition to adulthood and becoming full members of the society. These can involve apparently unbearable levels of agony – yet all members of the community undergo these rituals, often with what looks like stoic resilience to outsiders. It is likely that psychological factors such as expectation and group participation help them to endure the experience.

Hypnosis is thought to be a state of focused attention in which a subject is particularly susceptible to suggestion. Scanning by fMRI reveals that, under hypnosis, the brain responds to suggested stimuli in the same way as the non-hypnotized brain responds to comparable actual stimuli. For instance, a 2013 study reported that the same areas of the brain were activated to the same degree when hypnotized subjects were told they were receiving a painful stimulus as when they received it when not hypnotized. In the test, volunteers had a hot probe placed on their hand and subsequently recorded a pain rating of 5. When they were hypnotized and told the probe was again being activated – even though it was not – they recorded the same response. The brain scans were a close match for the genuine stimulus.

Pain, which seems to be the clearest and most intrusive of physical sensations, is not all we intuitively feel it is. The brain can conjure it with no reason or dispel it despite massive trauma; it represents the ultimate demonstration of mind over matter.

PHYSICAL AND EMOTIONAL PAIN ARE COMPARABLE

A study in 2013 found that the brain responds to emotional pain by releasing painkilling opioids in much the same way as it responds to physical pain. While a PET scanner observed their brains, subjects were told they had been rejected by potential partners they had expressed an interest in on an online dating site. Those subjects who expressed least distress had a higher release of opioids in their brains. The areas of the brain activated were the same as those activated by physical pain.

CHAPTER 8

Lessons from LESIONS

'Brain surgery is a terrible profession. If I did not feel it will become different in my lifetime, I should hate it.'

Wilder Penfield, 1921

There are many types of neurological illness or damage and their study has played a vital role in the unfolding story of neuroscience. Clinical investigations, attempted treatments and autopsies have all yielded new information about the brain and nerves. Scientists have not always been scrupulous in putting their patients' best interests first, though.

People have attempted to cure madness with brain surgery – however brutal and elementary – for thousands of years. In this painting from 1494, The Extraction of the Stone of Madness, *Hieronymous Bosch depicts an early operation.*

St Nilus heals a possessed boy by anointing him with oil drawn from a lamp burning in front of a picture of the Virgin Mary.

The 'sacred disease'

Perhaps the most fruitful neurological disease for science has been epilepsy. It is one of the earliest recorded medical conditions. In Ancient Greece, it was known as the 'sacred disease'. Around 400BC, Hippocrates remonstrated against the notion that epilepsy was the result of a visitation by the gods or divine retribution. He wanted it to be treated as an illness like any other, requiring medical treatment rather than charms and incantations peddled by charlatans. Unfortunately, his measured attitude did not prevail and demonic possession and other superstitious causes were commonly cited until at least the end of the 17th century.

Even when epilepsy was regarded as a physical ill rather than a divine punishment, there was no consensus regarding cause, and little effective treatment. One concoction used to treat epilepsy included powdered human skull, mistletoe and peony roots, and seeds gathered under a waning moon. Bloodletting was also popular (as it was for many ills). Galen thought epilepsy occurred when the ventricles became blocked with phlegm; in the 16th century, Paracelsus thought seizures were the result of the vital spirits boiling over in the brain; and Thomas Willis blamed explosions in the animal spirits in the *sensorium commune*.

Epilepsy is not insanity

> **ELECTRICAL BRAINSTORMS**
>
> Epilepsy is a medical condition characterized by seizures caused by sudden bursts of disorganized electrical activity within the brain. They interrupt normal brain functioning and can include loss of awareness, convulsions, stiffening of the limbs, jerky movements or unusual sensations.

It was against this background of superstition that Jean-Martin Charcot (1825–93) encountered patients with epilepsy shut up alongside the mentally ill and criminally insane in the Salpêtrière, Paris. He separated out a ward of women he diagnosed as not insane, but suffering from 'hystero-epilepsy'. During the 19th century there was considerable effort to describe the course and nature of seizures of different types, even if they were not understood.

The physiologist Marshall Hall (1790–1857) proposed the first physiological explanation for epilepsy. In 1838, he suggested that heightened activity in part of the reflex arc triggered the problem, suggesting the dysfunction lay in either the sensory or central connecting part of the arc within the spinal cord. He thought the neck muscles going into spasm prevented the flow of blood from the brain, causing congestion of the brain. This led to loss of consciousness, and spasms of the larynx which then caused convulsions. He advocated tracheotomy to prevent the convulsions.

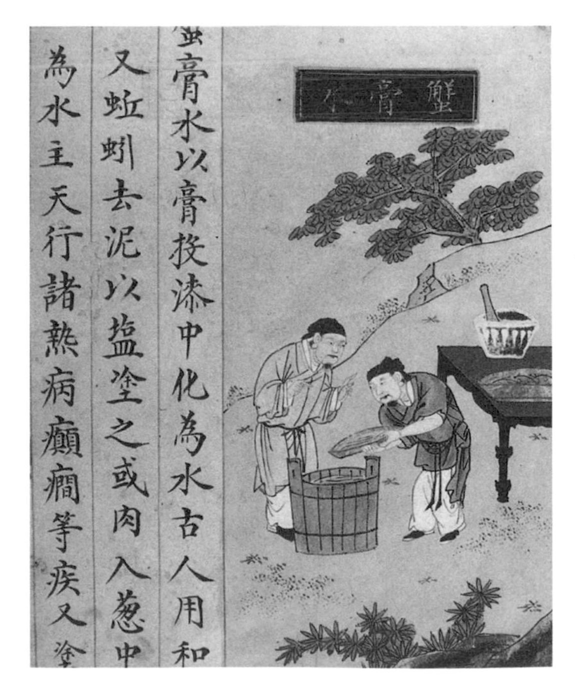

A Chinese illustration of the preparation of crab-roe water, used to treat epilepsy.

Sparking seizures

The first suggestion that electrical discharge might be implicated in epilepsy came from Robert Bentley Todd in 1849, but the discovery is usually credited to John Hughlings Jackson in 1873. Jackson defined epilepsy as the 'occasional, sudden, excessive, rapid, and local [electrical] discharge of grey matter'. He observed his patients (and his own wife) during seizures and noted their progression. He named a type of simple partial seizure, the Jacksonian seizure (or 'Jacksonian march'), which begins with tingling, twitching or weakness

in a finger, toe or one side of the mouth and then extends to the entire hand, foot or facial muscles. It affects one side of the body only and the patient doesn't lose awareness.

Proof that electricity is involved only became possible with Hans Berger's invention of EEG in 1929 (see pages 171–4). Berger managed to show the eccentric patterns of electrical discharge which occur in the brain during an epileptic seizure, proving the problem is electrical and that it originates in the brain. The discovery was a mixed blessing for patients with epilepsy as for several decades they would be subjected to terrifying and often highly damaging experimental treatments.

Maladies of the mind

People with epilepsy were not the only ones to suffer the results of ignorance. Mental illness has long been attributed to the influence of demons or evil spirits. Treatment has often been directed at driving out the recalcitrant spirits and included beating, shackling in leg-irons and even starvation. By contrast, Hippocrates proposed that mental illness was caused by an imbalance of humours; he aimed to treat it by rebalancing them.

These two opposing views, that mental illness is a physical or a spiritual ailment, co-existed for centuries. They produced a strange mix of treatments ranging from the harshest forms of savagery to behaviour therapies. There was no specific target for any of these practices; they treated the whole body (or the spirit) and recognized no locus of disease.

Head cases

The notion that mental illness might be a disease of the brain only emerged in the 19th century with the advent of general anaesthesia and effective antiseptics. These made surgery much safer, and as a result psychosurgery took an experimental turn.

The early neurosurgical treatments targeted the brain as the source of the dysfunction, rather than the whole body. But they were based on flimsy theories and therefore a dangerous hit-and-miss affair. The idea of just opening up the brain and digging around in it, cutting bits out, severing nerves and hoping for the best is

A woman diagnosed as suffering from mania.

the stuff of nightmares now. Yet that was the early history of psychosurgery.

The start of psychosurgery

How would you feel about someone chipping a hole in your skull with a stone without anaesthesia, while you were fully conscious? Unenthusiastic, probably. Yet the earliest form of surgical intervention is trephining, or trepanning – making a hole in the head. Evidence of trephined skulls, with healing around the cut edges of the bone, dates back to Neolithic times and the procedure has probably been carried out somewhere in the world ever since.

You don't need all that brain

For thousands of years, if someone had a profoundly damaged limb, amputation was regarded as the best or only option. It seems reckless to take the same approach to the brain, but early interventions did indeed involve removing or destroying entire parts of this organ.

The first attempt at psychosurgery, aside from trephination, was carried out in Switzerland in 1888. The doctor who performed the operations, Gottlieb Burckhardt, was not a surgeon but a psychiatrist and the director of a small mental hospital. He carried out experimental surgery on six patients, removing parts of their brains in an attempt to reduce the symptoms they suffered as a result of various forms of mental illness. In modern terms, his patients could variously be described as having schizophrenia, mania and dementia. Between them, they suffered from auditory hallucinations, paranoid delusions, aggression, agitation and violence. Burckhardt believed mental illness was caused by physical problems in the brain and aimed to alleviate symptoms by removing the damaged bits. Unfortunately, he had no sound way of telling which parts of the brain might be causing the problems.

James Norris, a patient in Bedlam, was shackled and isolated for ten years. His plight, revealed in 1814, prompted legislation to regulate conditions in madhouses.

The results were not encouraging: of his six patients, one died following epileptic convulsions five days later, one committed suicide, two showed no improvement in their condition and two became calmer. Only two had no side effects, which otherwise ranged from epilepsy through language difficulties and problems with

A HOLE IN THE HEAD?

The discovery of the first Neolithic trephined skulls was met with incredulity. In 1865, the explorer and ethnologist Ephraim Squier received the gift of a skull from an Inca burial ground near Cuzco. It had a square hole cut in it, with sides of around 1cm (½in). Squier concluded that the hole had been cut deliberately while the skull's owner was living – and that the patient had survived the operation.

The New York Academy of Medicine refused to believe that a 'primitive' Peruvian Indian could carry out such an operation and have their patient live, particularly as the survival rate of trephination patients in the 1860s was around 10 per cent. But high infection rates in 19th century hospitals made the operation more dangerous than if it had been carried out in a cave. Recent estimates suggest that survival rates in the distant past could have been 50 or even 90 per cent. Also, the operation was only used in the 19th century in cases of severe head injury (so the patient was quite likely to die with or without the intervention) while it was almost certainly used for lesser conditions in the past.

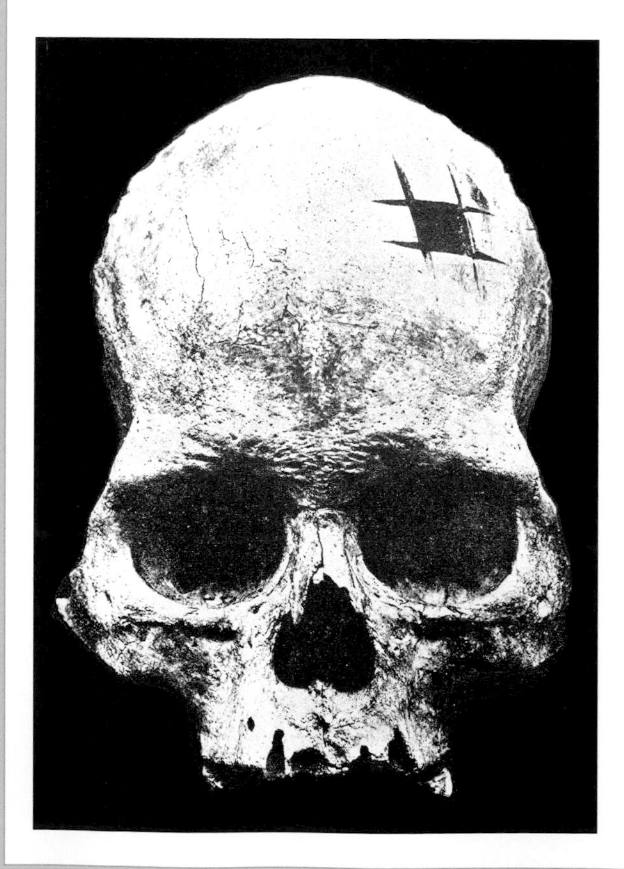

Many of the societies that have used trephining have been pre-literate so we have no record of why the operation was carried out. In Europe, trephining was used to treat epilepsy and some forms of mental illness, from the time of Aretaeus the Cappadocian (*c.*150AD) until the 18th or even 19th century. It was supposed to allow evil vapours or humours to escape from the brain.

movement. His report of the proceedings met a hostile response and he gave up psychosurgery – a fortunate outcome for his remaining patients. Psychosurgery was not tried again until the 1930s. But when it came back, it was bold and brutal.

A brain with two halves

In 1928, surgeon Walter Dandy removed the entire right hemisphere (an operation called hemispherectomy) in an attempt to treat patients with otherwise inoperable and terminal tumours (called glioma). Of five patients, three died within three months (one immediately). In 1933, another surgeon, W. James Gardner, operated on a further three patients with epilepsy and two years later one was seizure-free, cognitively well and able to walk, with no recurrence. Although hemispherectomy was soon superseded as a treatment for glioma, it was used for epilepsy, first by the South African neurosurgeon Roland Krynauw in 1950. His apparent success in curing seizures in young patients led to the enthusiastic adoption of the procedure.

Unfortunately, problems such as haemorrhage and encephalitis manifested later, even years after surgery. Surgeons tried different variations of hemispherectomy, leaving some tissue in place but disconnected from the corpus callosum, the thick band of nerve fibres which links the left and right hemispheres and allows communication between them.

Sperry and the 'split brain'

Severing the corpus callosum proved an effective treatment for severe forms of epilepsy in the 1960s. It was slightly less extreme than removing an entire hemisphere. It means the two halves of the brain can function independently, but cannot communicate, so the random triggering of nerves in the epileptic seizure cannot spread from one hemisphere to the other. As well as curing epilepsy, the operation provided new insights for neuroscience. American neuropsychologist Roger Sperry (1913–94) studied eleven patients with 'split brain' to investigate how the

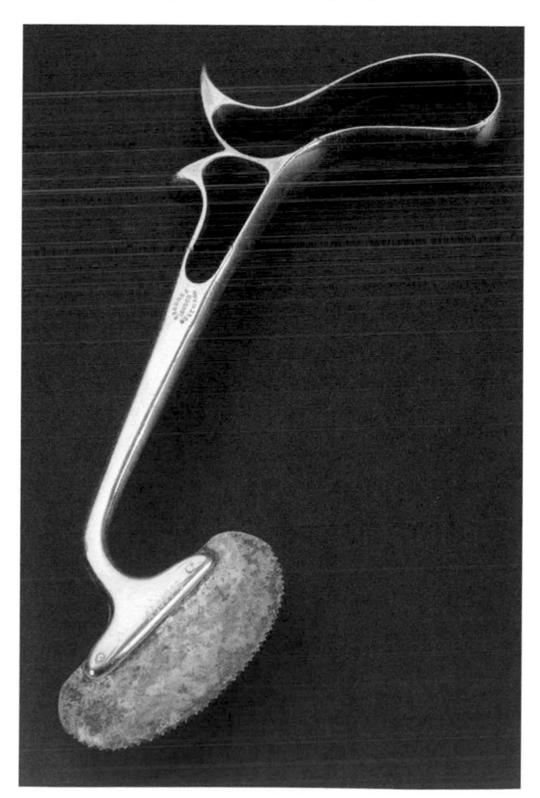

This special saw was used to cut open the skull to access the brain. It was invented by Sir Victor Horsley (1857–1916), an English surgeon and physiologist.

hemispheres usually operate together.

Remembering that the main centre for speech is in the left hemisphere and that input from and control of the left side of the body are handled by the right hemisphere, Sperry was able to demonstrate that the hemispheres act together to articulate information relating to the left side. For instance, if an image was shown to the left eye only, the patient couldn't say what he or she had seen: there was no way for the visual centre and the speech centre to communicate with each other.

Similarly, if a patient felt a texture or substance with their left hand, they could not say what it was. In normal activity, we can use both eyes, both ears, both hands and so on, so these limitations were not immediately apparent, but they were clearly demonstrated in the experimental setting. Although the right side of the brain couldn't articulate in speech or writing what an object was, the patient was able to draw it or identify a similar object. The right side was also better at visual-spatial tasks. Sperry further found that if an object was shown to one eye and then the other, the patient had no recollection of having seen it before – each hemisphere seemed to form its own memories. Sperry won a Nobel prize for his work on split-brain patients.

Roger Sperry receiving his Nobel prize for work on split-brain patients, 1981.

The most famous brain operation

No sooner had Walter Dandy begun removing entire hemispheres of the brain than another surgeon hit upon the idea of annihilating specific parts of it. This uninformed and blasé approach to brain surgery of the mid-20th century included the notorious pre-frontal lobotomy, using

spikes to destroy the connection between the frontal lobes and the rest of the brain, which was little more than brain butchery.

Calm monkeys, bad idea

In 1935, the Portuguese neurologist António Egas Moniz supervised the first leucotomy, or lobotomy, in a hospital in Lisbon. He was not a neurosurgeon, and his hands were severely disabled by gout, so his assistant Pedro Almeida Lima carried out the procedure under Moniz's direction. The thinking behind it was that the frontal lobes of the brain are implicated in many disorders and damage to them through illness or accident often results in changes of personality or behaviour.

In 1935, the American neurologist John Fulton had displayed two chimpanzees previously noted for their challenging behaviour but which, after a full lobotomy, had become much calmer and seemed happier. Moniz then decided to venture trying the technique on human patients. He explained the neurological theory behind the operation in terms of association: the brain has formed fixed but unhealthy associations (neural pathways) which lead to obsessive ideas. He argued that these could be broken only by physically destroying the connecting pathways in the brain. He thought the brain would adapt functionally, building new pathways which could be healthier.

Moniz and Lima's first operations involved drilling holes in the patient's skull and injecting ethanol into the white matter of the frontal lobes to destroy the fibres that connected them to other areas of the brain. After the eighth operation, and frustrated by the sometimes limited success of their procedure, they resorted to inserting an 8cm-long needle into the hole, and wiggling it around to destroy the nerve connections. Moniz received the Nobel Prize for medicine for this work in

LEFT-BRAIN, RIGHT-BRAIN

Sperry's work helped to prompt a rush of pop-psychology articles claiming that the left side of the brain is analytical and logical and the right side of the brain is creative and imaginative. This idea was expanded to persuade people that one side of their brain is dominant and this determines whether they are better at being logical or creative. But this doesn't follow on from Sperry's work and it's not supported by neuroscience. Studies using fMRI to watch the brain working show that people do not preferentially use one side of the brain over the other.

The psychologist Robert Ornstein suggested in 1970 that people in industrialized western societies only use half of their brain properly – we focus so much on logic, language and analysis that we are out of touch with our intuitive side. This theory came to affect educational practice, with some critics claiming our teaching style favours 'left-brain' learners. But again, there is no evidence for different thinking styles related to the hemispheres. Even so, this type of misconception became hugely popular and is still widely believed.

1949, an award which has invited some controversy.

(Ice) pick and choose

Lobotomy quickly became very popular. In the USA, psychiatrist Walter Freeman carried out his first lobotomy (he renamed the operation) on Alice Hood Hammatt, a 63-year-old woman from Kansas. He believed mental illness was caused by an 'overload' of emotions and aimed to cut nerves in the brain to reduce the emotional load and calm his patients. He became not only an enthusiastic and prolific lobotomist, but something of a showman.

Freeman developed a new technique which did not require drilling into the head. Called the 'ice-pick lobotomy', it was every bit as terrifying as it sounds. He used general anaesthesia or electroshock to render his patient unconscious, then inserted a spike like an ice pick into the eye socket above the eyeball and used a mallet to drive it into the brain. He moved the spike at carefully prescribed angles and depths, destroying connections to the frontal lobe. Then he did the same through the other eye socket. Sometimes, for dramatic effect, he did both eyes at once. He performed around 2,500 lobotomies in his life, sometimes carrying out 25 a day, each taking just ten minutes. He was eventually banned from performing lobotomies in 1967 after a woman he had lobotomized for the third time suffered a brain haemorrhage and died.

Between 40,000 and 50,000 lobotomies were performed in the USA, and 17,000 in the UK, mostly during the 1940s and 1950s. Finland, Norway and Sweden together

Egas Moniz pioneered lobotomy, a procedure that would damage thousands of patients over a few decades.

carried out around 9,300 – a higher per capita rate than the USA. The operation was used to treat various types of mental illness including schizophrenia and depression, but sometimes it was used on children labelled as 'difficult' and even to reduce chronic pain. Eva Perón, the wife of the Argentinian president Juan Perón, was lobotomized to control the pain of her cancer.

Lobotomy was effective in around a third of cases. People agreed to it because they were desperate. Mental hospitals were full of strait-jacketed, subdued or ranting patients who had little hope of recovery or effective treatment. Lobotomy seemed to offer some hope of escape from what

would otherwise be a lifetime in the asylum. Among those who survived relatively intact, dulling of the personality and lethargy were common consequences.

Lobotomy fell out of favour in the 1950s when psychoactive drugs became available. Inevitably there remained many patients permanently damaged by lobotomy who might otherwise have been helped by new treatments. The USSR was the first country to ban lobotomy, declaring it 'contrary to the principles of humanity'. In 1977, the US Congress set up a body to investigate claims that lobotomy had been used alongside other psychosurgical techniques to subdue and control minority groups.

Shocking treatment

The realization that epileptic seizures are caused by bursts of electrical activity in the brain also presented the possibility of using electricity on the brain therapeutically. Another of the bold, optimistic projects of the 1930s, electroconvulsive therapy – or electric shock treatment – deliberately precipitated a seizure to 'jumpstart' the brain. It was used to treat a variety of mental illnesses, especially severe depression.

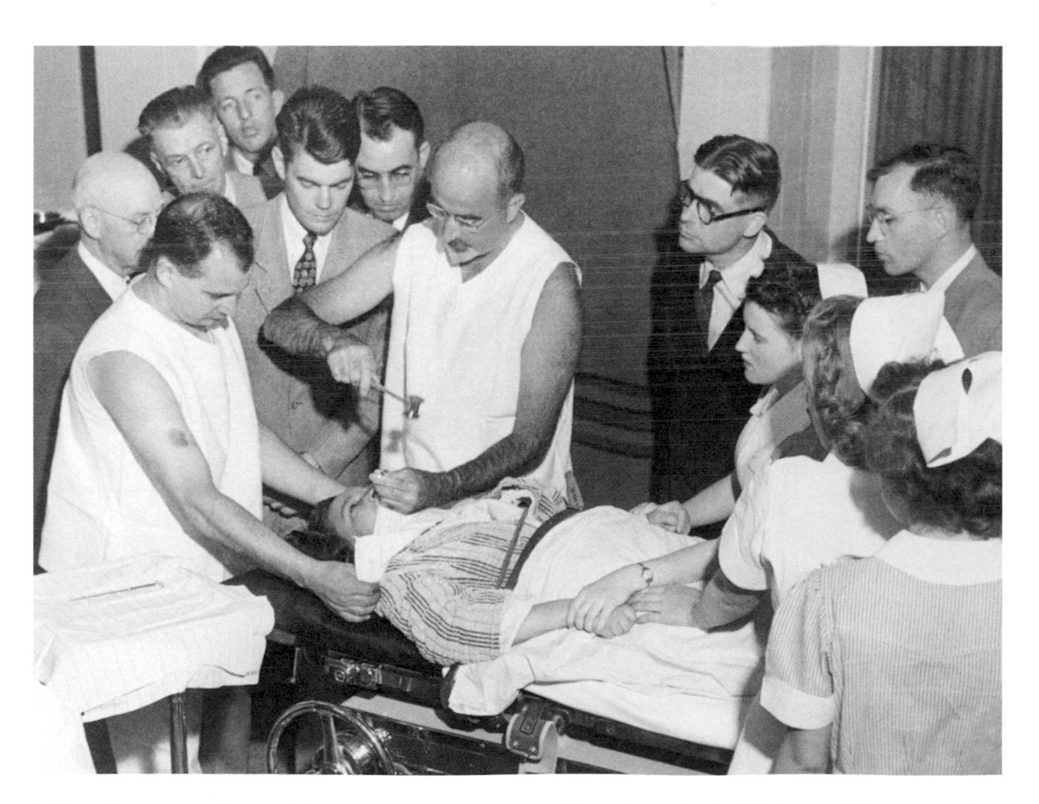

Walter Freeman performs a lobotomy using an instrument like an ice pick which he invented for the procedure. Inserting the instrument under the upper eyelid of the patient, Freeman cuts nerve connections in the front part of the brain.

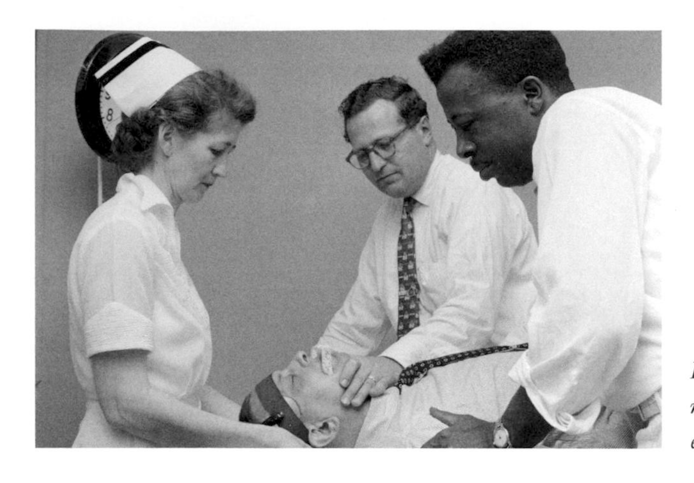

In 1955, a patient at a mental hospital undergoes electroconvulsive therapy (ECT).

Fits and starts

Hippocrates was the first to notice that convulsions sometimes seem to cure mental illness. He observed patients whose mental health had improved after malarial fever had induced convulsions. Others noted the same over the centuries, and it became commonly believed that people who had epilepsy could not also be mad (in the terminology of the time). Nevertheless, epileptics continued to be locked up in asylums.

From 1917, doctors tried instigating convulsions in the hope of curing mental illness. The first cases involved giving patients blood infected with malaria. Then, in 1927, the Polish neurophysiologist Manfred Sakel discovered that by giving a patient a massive dose of insulin he could cure her mental illness. Insulin is a hormone created by the body to regulate the level of sugar in the blood. Too much insulin reduces blood sugar levels and triggers coma and convulsions. From 1930, Sakel perfected insulin shock as a treatment for schizophrenia. In 1933, a Hungarian physician named Ladislaus von Meduna found that by injecting the drug metrazol he could induce severe convulsions which treated mental illness. Unfortunately, the terrifying convulsions were so severe that nearly half his patients suffered spinal fractures. In 1940, the treatment was modified by adding curare (see page 103) to reduce the convulsions, and later an anaesthetic was added so that patients were not conscious during treatment.

> *'[P]refrontal lobotomy . . . has recently been having a certain vogue, probably not unconnected with the fact that it makes the custodial care of many patients easier. Let me remark in passing that killing them makes their custodial care still easier.'*
>
> Norbert Wiener, American philosopher and mathematician, 1948

A safer treatment?

ECT was first developed by Italian physicians Ugo Cerletti and Lucio Bini in 1937 as a safer and more pleasant alternative to the drug metrazol. They developed the

technique of applying short-duration electric shocks to the brain, first with animals and then with schizophrenic patients. They found that as the shock also caused retrograde amnesia, the patients had no memory of the treatment so were unafraid of it. After 10–20 shocks, applied on alternate days, there was a startling improvement.

Opium poppies are among the sources of mind-altering drugs that have been used since prehistory.

ECT was soon widely used in mental hospitals around the world. But many abused it, using it to subdue or control patients rather than to treat them. This abuse was widely publicized in the 1962 novel *One Flew Over the Cuckoo's Nest* (followed by the Academy Award winning film in 1975), and the tide turned. Following complaints and prosecutions, the therapy fell out of favour and was soon replaced by new drug therapies. It has since been revived, with better procedures and safeguards. But although it's effective, we don't really know how or why it works.

Keep taking the pills

The use of drugs (usually plant extracts) which have an effect on the brain is at least as old as trephining. They have been used to produce states of trance or frenzy for religious or shamanic activity, for some level of anaesthesia or pain relief and to treat mental illness. Substances derived from poppies (opiates), the coca plant (cocaine), alcohol and tobacco are just some of the many naturally occurring mind-altering drugs which have been with us for thousands of years.

Drugs and talking take over

Today a combination of drugs (as a physical remedy) and talking therapies (as a psychological remedy) has replaced most psychosurgery and shock therapies. A patient might be offered both a talking therapy, such as cognitive behavioural therapy (CBT), and medication, the one treating the mind and the other changing the physical–chemical state of the brain. Modern neuroscience would suggest that talking therapy can also change the physical and chemical state of the brain, or at least its neural wiring – which is manifest in physical and chemical states at the cellular level.

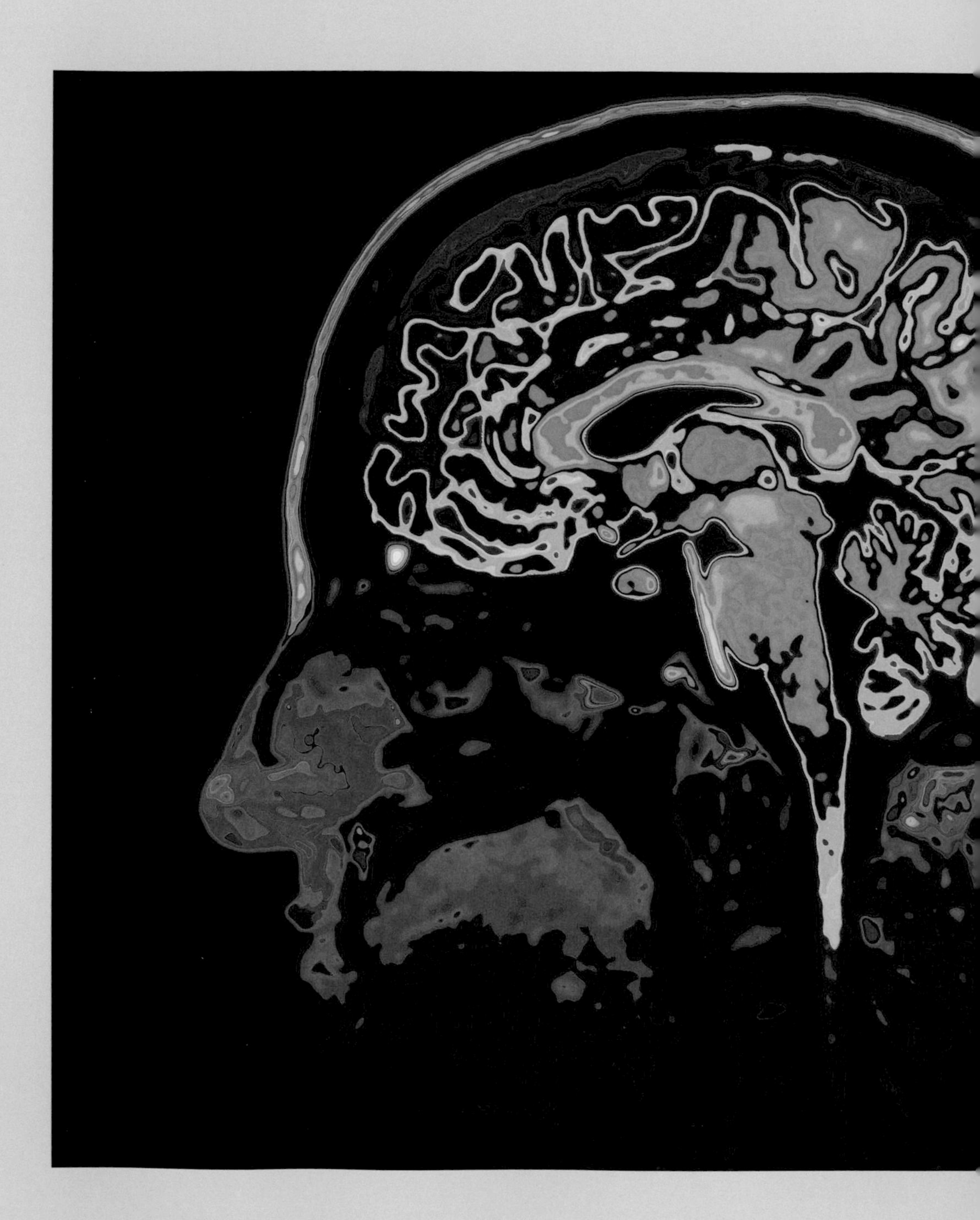

CHAPTER 9

What's going on IN THERE?

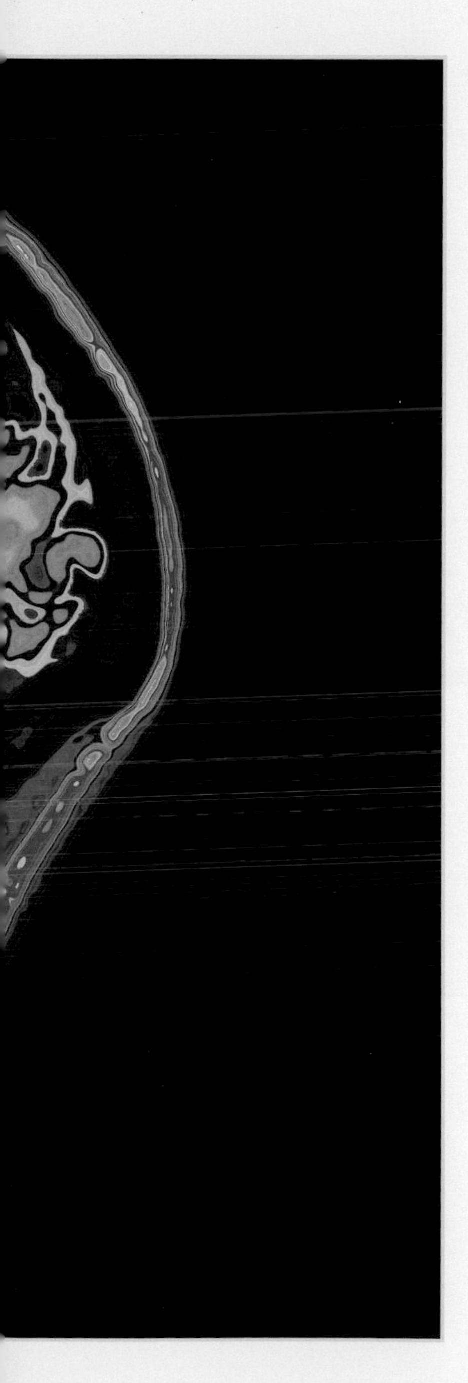

'Think mind reading contrary to common sense, wise provision of the Bon Dieu that we cannot read each others minds, twould stop civilization and everyone would take to the woods.'

Thomas Edison, 1885

Until the 20th century, there were only two ways to localize brain function. One was to look at damaged areas of the brain and compare them with impairments in brain function. The other was to expose the brain and basically poke it about – even to the extent of cutting or destroying great chunks – and observe the effects. Brain imaging technologies have changed all that. Now we can see inside the working brain, watching different areas fire up as a subject acts or thinks, without causing any damage at all.

MRI and other scanning technologies enable us to watch the brain as it thinks, dreams and does its work controlling the body.

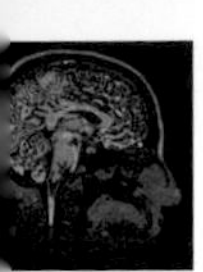

Damage and limitations

There are two good reasons for wanting to look at the living brain. One is to find out more about how it works and extend our knowledge of localization. The other, specific to an individual patient, is to diagnose problems and perhaps offer treatment.

Up until the end of the 19th century, it was usually necessary for a patient to be dead before working out what was wrong with his or her brain. If doctors kept a record of symptoms, they might be able at autopsy to correlate them with any lesions they found – or they might not. This approach produced some useful information about which part of the brain dealt with function, but tended to indicate general areas of activity rather than pinpointing precise locations. That's hardly surprising; an injury or disease doesn't 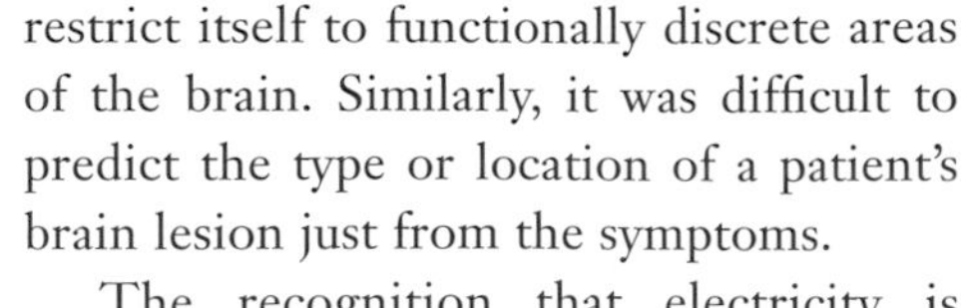restrict itself to functionally discrete areas of the brain. Similarly, it was difficult to predict the type or location of a patient's brain lesion just from the symptoms.

The recognition that electricity is involved in neural transmissions led investigators to the brain, brandishing their electrodes. Long before they tried adding more electricity with ECT, they began to monitor the electricity already in the brain.

A rabbit in the headlights

Richard Caton (1842–1926) was the first person to measure brain activity by recording its electrical activity. He had studied medicine under David Ferrier, and began with his findings. Ferrier had found that electric stimulation of parts of the motor cortex caused a dog or rabbit to move particular body parts; destroying those areas of the cortex caused paralysis in the corresponding body parts. Caton decided instead to measure the electric current in the brain when the animal was responding or acting.

Fixing electrodes to the part of the rabbit's brain which Ferrier had located as responsible for moving the eyelids, Caton found that shining a bright light in the rabbit's eyes caused an electrical current in that part of the brain. It was a revolutionary discovery – the first evidence of spontaneous electrical activity in the living brain. Caton wired up his subjects (animals) and allowed them to walk, eat and drink while he monitored their brain activity to find out which parts were involved in common activities.

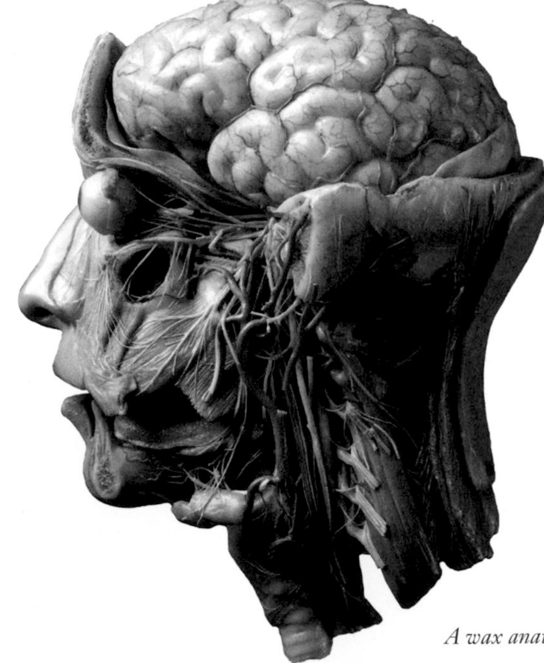

A wax anatomical model of the head, made in Europe in the 1900s.

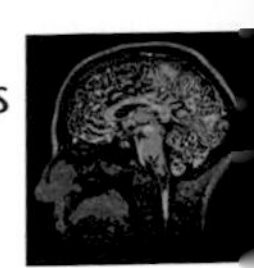

The next great breakthrough came in 1929, when German psychiatrist Hans Berger published his findings relating to the electric patterns generated by the human brain. He had recorded these with his new invention, the electroencephalograph (EEG). The path to the EEG was a long one. It began in the 1890s with two unrelated incidents: an accident that befell Berger, and the work of two psychiatrists who wanted to explain the physics of the soul.

The energy of the soul

One of the most important and influential discoveries in physics in the 19th century was the conservation of energy. There were attempts to apply the principle in different fields, including neuroscience. The German neuropsychiatrist Theodor Meynert wanted to establish a psychophysiology that bridged the mind/body or body/soul gap and looked for a physiological conservation of energy. He stated that when energy is produced in one part of the brain, triggering a thought or action, an equal quantity of energy must be lost elsewhere in the brain – otherwise the human soul would violate the laws of physics.

Theodor Meynert tried to apply physics to the brain's energy economy.

Meynert and his colleague Alfred Lehmann proposed that the flow of blood and the flow of energy are connected. By restricting blood flow to one part of the cerebral cortex, extra blood (and therefore extra energy) can be supplied to other areas. The chemical energy (ultimately derived from food) that the brain uses is converted into other forms of energy as the brain works. This can take three forms: heat, electricity and something they called 'P-energy' – the psychic energy associated with different mental states. To produce a quantum of P-energy, an equivalent quantum of some other type of energy must be converted.

While Meynert and Lehmann were exploring the potential physics of psychic energy, a young soldier was putting his own psychic energy to good use.

A close call with destiny

The young Hans Berger went to university to train as an astronomer. But it did not suit him, so he left in 1892 and took a military commission. He then had an accident which changed the course of his life. Berger was thrown from his horse in

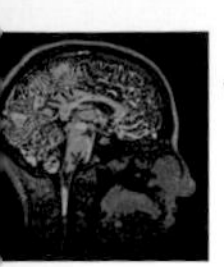

Hans Berger came to neuroscience because of a psychic experience he had undergone as a young man.

front of a moving gun carriage. Fortunately, the vehicle stopped just in time, saving him from certain death.

That evening, Berger received a telegram from his father asking after his safety. It transpired his sister had felt such an overwhelming sense of dread at the moment of the accident that she had insisted their father write and check Hans was safe. Berger became convinced that his own terror must have been communicated telepathically to his sister. When his year of military service came to an end Berger returned to university, but this time to study medicine. Once qualified, he began to work in psychiatry.

Then Berger came across the work of Lehmann and Meynert. Here was something he felt might offer a physiological explanation for his psychic experience. He set about trying to untangle the electrical wiring of the brain in the hope of understanding psychic energy.

Blood and brains

For 30 years, Berger meticulously investigated the supply of metabolic energy to the brain and its conversion to heat, electricity and P-energy, or mental phenomena. By day, he led a very structured life, teaching students and running his research environment in line with strict rules and routines. But privately he conducted research into perhaps the most borderless territory there is – the production of emotions, thoughts and mental states. Lehmann had suggested that the consequences of research into this area were 'utterly unforeseeable'.

Berger began by measuring blood flow to the living brain. No one had done this directly before, but Berger had a ready supply of patients who had endured craniotomies (removal of part of the skull). One was a young factory worker who had been left with an 8cm hole in his cranium following two operations to try to remove a bullet from his head. The young man agreed to be Berger's experimental subject.

Berger made a rubber cap filled with water which he attached to the hole in the man's skull. He connected this to an instrument that recorded changes in pressure. He also measured blood pressure in the man's arm. He then subjected him to unpleasant shocks and pleasant experiences and asked him to carry out mental tasks. By comparing the changes in blood flow in the arm and blood flow in the brain, Berger discovered that cerebral blood flow increased with pleasant sensations and decreased with unpleasant ones, confirming a proposal made by Lehmann and Meynert. It was an

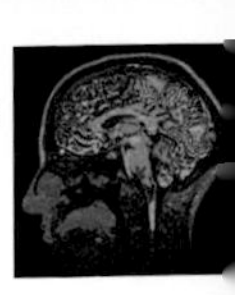

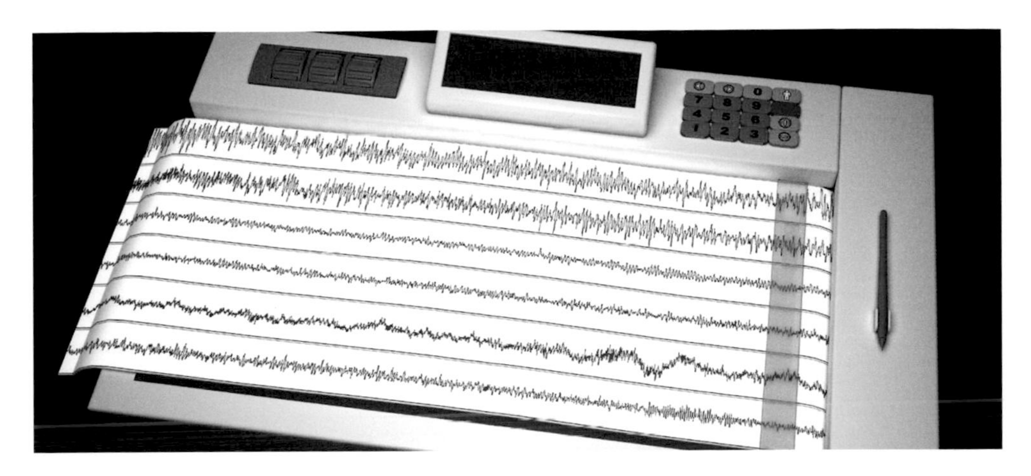

A modern EEG trace, showing patterns of brain activity.

interesting start, but it did not get to the heart of the matter.

Measuring psychic energy

Berger turned instead to measuring electrical current in the brain, reasoning that if he could measure the energy being produced by the brain and then exclude the energy converted to electricity and heat, he would be left with a measure of psychic energy.

He spent many fruitless and frustrating years trying different pieces of equipment and refining his methods, sometimes distracted or removed from the task by personal or professional obligations (and by World War I). In 1910 he bought a galvanometer – a device that measures small changes in electrical current. Working with this as a basis, Berger had developed his first electroencephalogram (EEG) by 1924.

EEG at last

Berger's first success was with a 17-year-old student called Zedel who had been left with a large hole in his skull after surgery to remove a tumour. Berger's recording was very basic and showed none of the peaks and waves that later EEGs would produce, but it was proof that the concept worked and this spurred him on to further efforts. He submitted his first paper on the human EEG in 1929. He had by this time made hundreds of recordings of normal and damaged brains, and soon identified the alpha waves that correlate with mental activity and the smaller beta waves that correspond to metabolic activity in the cortex. Berger had done all the work alone and in secret, never revealing even to his closest academic colleagues what he was doing in his laboratory.

EEG was, and still is, used to diagnose epilepsy and for investigating brain tumours and degenerative brain disease. It can also determine whether brain death has occurred and monitor anaesthesia, particularly during a medically induced coma.

Between around 1935 and 1970, EEG revolutionized neurology and for many

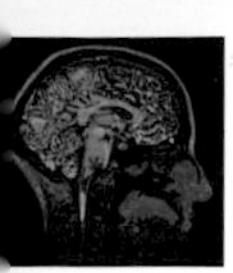

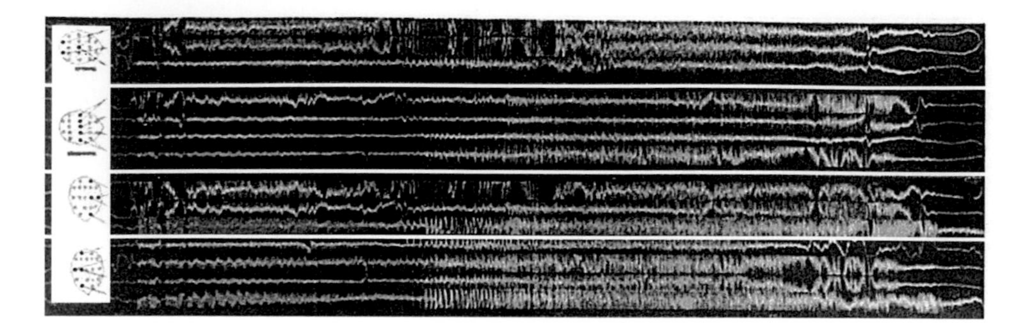

The onset of an epileptic seizure is marked in an EEG trace by a sudden burst of electrical activity.

years was the only way of tracking brain activity. In 1895, German physicist Wilhelm Röntgen had discovered X-rays, which were also used to look at the static structure of the brain and reveal lesions, but they could show nothing of what the brain was actually doing. Nevertheless, both EEG amd X-rays meant that for the first time ever we could look inside the body without opening it up. Although EEG remains an important tool today, it has been replaced by more complex brain imaging methods in some areas.

Air heads

The first breakthrough using X-rays for neuroscience came in 1918 when Walter Dandy developed ventriculography. This involved introducing air into the ventricles of the brain through holes in the cranium. Dandy used the technique to diagnose hydrocephalus (excess fluid in the brain). As cerebrospinal fluid (CSF) and brain tissue look similar on X-rays it was difficult to detect excess fluid, but air in the ventricles shows up clearly. In the normal, unobstructed brain, the air dissipates over a few hours, but if a patient has hydrocephalus it takes much longer because the routes by which the air (and CSF) would dissipate are blocked.

The following year, Dandy introduced a variant called pneumoencephalography. This involved draining the CSF

A nurse squeezes gel under a rubber cap before carrying out an EEG procedure on a young patient.

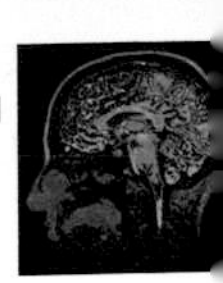

THE BRAIN AS MAGNET

Electrical current creates a magnetic field, and that happens in the brain, too. Magnetoencephalography (MEG) investigates brain functioning by measuring the very tiny patterns of magnetic field created by the brain's electrical activity. The magnetic field produced is around a millionth of the amplitude of background magnetic field in the urban environment, and can only be measured by very sensitive equipment. MEG only became possible with the invention of the superconducting quantum interference device (SQUID) in the 1960s and was developed to a useful standard in the 1980s. Around 50,000 neurons must be active to produce a measurable field, and only activity on the surface of the cortex can be detected at present. MEG is used in neurological research, pinpointing precise locations of brain activity, often used in combination with fMRI (see page 177). Its potential as a diagnostic tool is also being explored.

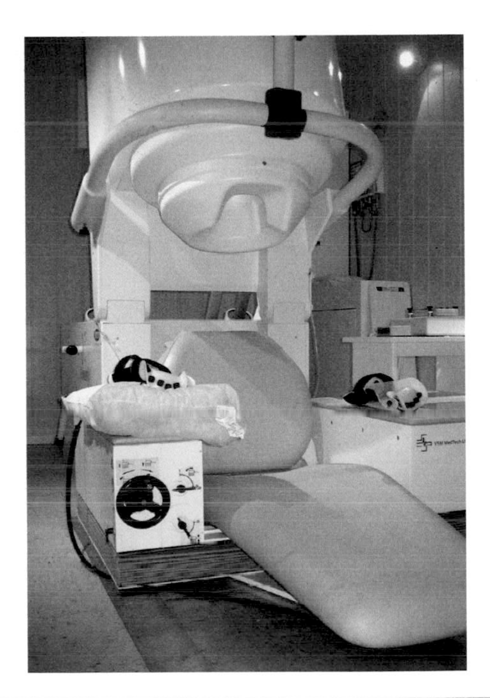

The instrument used for MEG can be used with a sleeping or even moving patient, making it more versatile than fMRI.

completely using lumbar puncture (a needle inserted into the spine) and replacing it with air, oxygen or helium. Although the procedure was both dangerous and very painful, it allowed the structure of the brain to show up more clearly. Even so, only a lesion right on the edge of a cavity, or so large that it distorted adjacent areas of the brain, was visible. The unpleasantness of the procedure meant it was rarely repeated to track the progress of a lesion. Fortunately, it was rendered obsolete in the late 1970s by more accurate, safer and more comfortable procedures.

Slices of brain

It took the development of computers in the 1960s for X-rays to be able to produce detailed images of the brain. From the indistinct, blurred resolution of

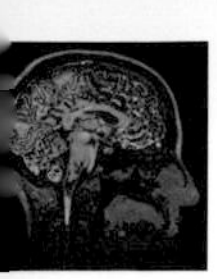

the early X-rays, fine, clearer structures began to emerge. Neurosurgeons could at last know what to expect before opening up a patient's cranium.

From fruit to brains

The breakthrough came with tomography, a technique for producing images like 'slices' through a solid object. From these slices it was possible to build up a three-dimensional model of the brain. The technology for doing this was first developed in the 1930s, but there were no computers available then to combine the data and create a composite image.

In 1959, American neurologist William Oldendorf watched a machine at work scanning fruit for quality control. Its job was to identify dehydrated areas of fruit that had been frostbitten. He was inspired to use the same type of technology to scan the human brain, effectively showing slices through the brain by scanning it with a beam of X-rays and building up a density map. Oldendorf built a prototype which used an X-ray source and a detector that rotated round a fixed object and could produce an X-ray picture from any angle.

It took a lot of development work, particularly in mathematics, before the first operational CAT (computer-assisted tomography) scanner was ready in 1971. Invented by British electrical engineer Godfrey Hounsfield, it took 160 parallel readings through 180 angles, each 1° apart. A scan took just over five minutes, with a further two-and-a-half hours to process the data. The first scan helped to diagnose a brain tumour in a 41-year-old patient which was then removed by a surgeon – a stunning success!

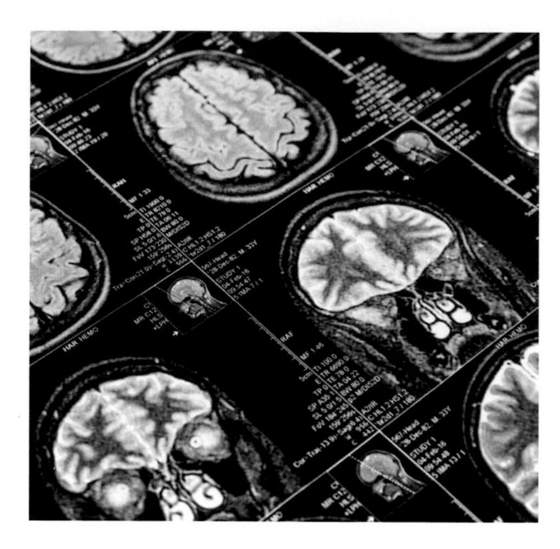

CT scans can show cross-sectional views through the brain at different angles.

The speed and resolution of CT (CAT) scanners has improved massively since the 1970s. It is now possible to scan hundreds of slices of the brain, each taking just a fraction of a second, and produce high-resolution images that allow detailed diagnosis.

CATs and PETs

CAT scans can show the structure of the brain, including indicating lesions and tumours, but cannot on their own show function or activity. About the same time that the CAT scanner was developed, positron-emission tomography (PET) also emerged.

Berger's linking of blood flow and neural activity has come to fruition in the PET scan, which looks at the metabolism of glucose in the brain, carried there by the blood, as an indicator of neural activity. A radioactively-tagged chemical (usually glucose) is injected into a subject or breathed in by them. The chemical has a short half-

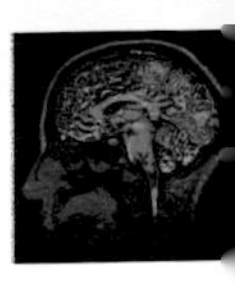

life, and as each radioactive atom decays it emits a positron and a neutron. When the positron encounters an electron, both are destroyed, releasing gamma rays. A gamma-ray detector in the scanner detects these emissions and produces an image from their concentration. Working on the assumption that metabolic activity correlates with use of glucose, PET scans can show patterns of activity in the brain on the basis of the concentration of tagged glucose that accumulates in the most active areas.

PET scans can also be used with other radioactively tagged chemicals to trace the concentrations of different neurotransmitters in the brain. Revealing metabolic activity and the release of neurotransmitters, PET scans let us watch the living brain in action. By combining a PET scan with a CAT scan, it's possible to superimpose levels of activity on a structural map of the brain, showing what is happening where.

The working brain

The next type of brain imaging to be developed was magnetic resonance imaging (MRI). This uses radio waves displaced by a magnetic field to produce a structural map of the brain. As it involves neither X-rays nor radioactive substances, it is considered safe for most people.

MRI is now most familiar in the form of fMRI (functional MRI) which reveals brain activity, pinpointing areas of the brain that 'light up' when a subject is subjected to a stimulus or carries out an activity. It was developed by the Japanese researcher Seiji Ogawa in 1990. fMRI relies on the assumption that increased blood flow corresponds to increased neural activity, and uses the difference in magnetization of oxygenated and deoxygenated blood to reveal activity in the brain or spinal column. It produces images of BOLD (blood-oxygen level dependent) contrast between areas of high and low activity. Unlike PET, fMRI can be used to monitor the brain over an extended period, so can be used while the subject carries out more complex tasks. The duration of PET scans is limited by the half-life of the radioactive agent.

From diagnosis to discovery

Although scanners are still used extensively in hospitals they have also enabled considerable research into the structure and functioning of the normal and abnormal brain. With the benefit of brain-imaging techniques, particularly fMRI, we can explore the localization of brain function in real time, watching mental activity as it happens.

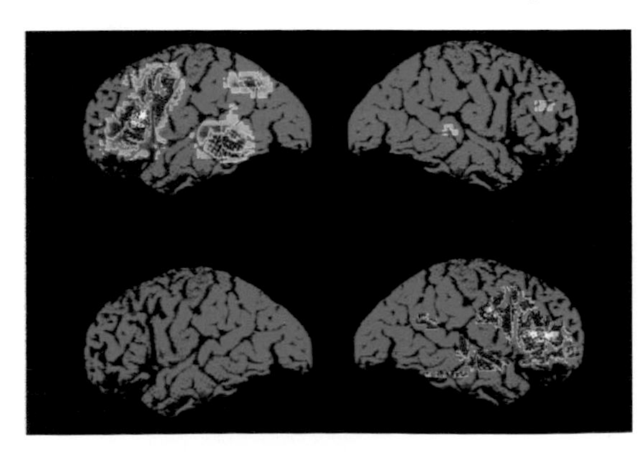

PET scans of a right-handed subject (top) and left-handed subject (bottom) as they carry out word-related tasks. The scan shows which areas of the brain are active, revealing opposite hemispheres operating in the two subjects.

CHAPTER 10

Thinking and BEING

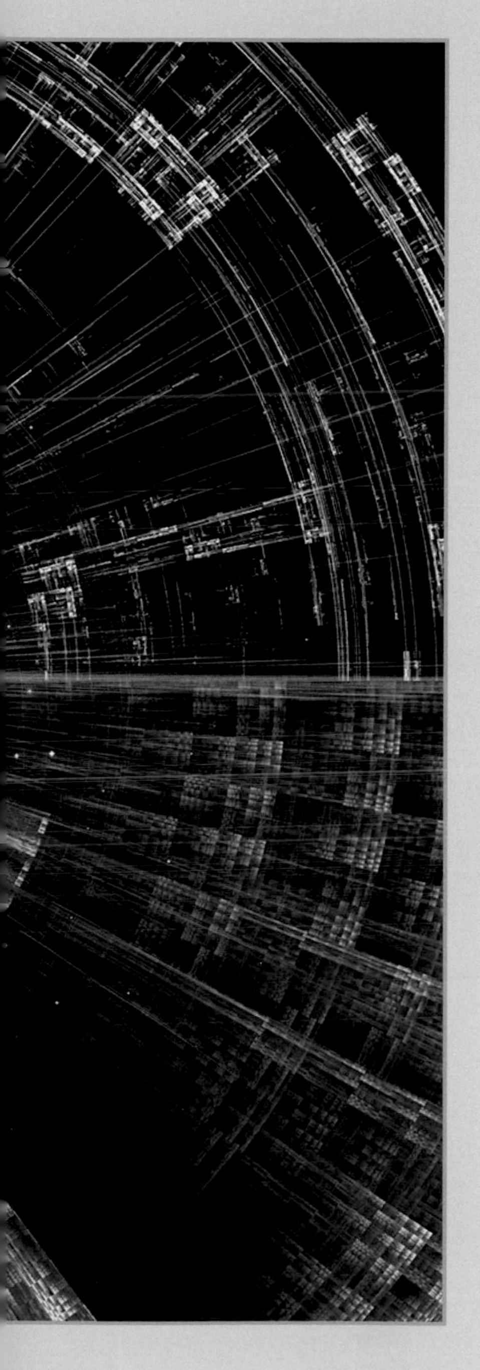

'To think and to be are one and the same.'

Parmenides,
late 6th/early 5th century BC

The sensory and motor tasks of the central nervous system are the easiest to investigate and localize, but perhaps its most interesting and most elusive function is mind: the emotional and cognitive aspects of the brain that define us as human and as individuals.

Most of us feel that our mind comprises our identity more than our body.

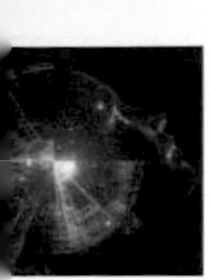

All in the mind

Investigating the activity of the sensory and motor nerves, while not easy, is comparatively easier than dealing with what goes on entirely inside the brain with no manifest signs in the rest of the body.

Purely mental activities include thought, memory, dreaming, imagination and creativity. Some of them can take place without the need for any external stimulus or interaction with the outside world. They are the most challenging events to study, and before the development of brain imaging technologies they were essentially hidden from research. Even the idea that they take place in the brain rather than elsewhere (or nowhere) in the body was hard to demonstrate. Yet it is the internal life that creates our sense of identity and our uniqueness. These aspects of brain activity are among the most fascinating and mysterious challenges for neuroscience and form the material of cognitive neuroscience.

Cognitive neuroscience

Cognitive neuroscience combines the disciplines of neuroscience, philosophy, psychology, linguistics, anthropology and artificial intelligence (AI). As such, only some of it is relevant to the story of neuroscience, but there are essential intersections. It

Producing artwork is apparently unique to humans and involves a host of cognitive abilities. The artist uses inspiration, imagination, memory, anticipation, critical appreciation and evaluation. And the brain handles the sensory and motor skills involved in actually producing a painting alongside the cognitive work.

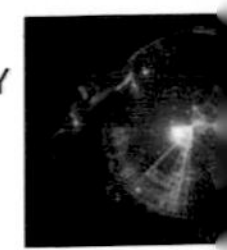

Seeing a single photograph can spark memories of a scene, a day, an event, or a person, recreating emotions from the past.

deals with topics such as memory, learning, language acquisition and processing, consciousness, perception and attention.

The first cognitive psychologists in the 1960s rejected the behaviourist approach to psychology (see page 185), which ignored anything taking place in the mind and focused only on the stimulus (input) and the resultant behaviour (output). Instead, cognitive psychologists aimed to show that perception is constructive: it starts with incoming information, but the brain works on this information to create something new, transforming it into meaningful perception and resultant action (or memory). The cognitive approach to behaviour relied on the idea that each act or item of incoming information is internally represented in the brain in patterns of neural activity. The cognitive psychologists took on the very part of the process that the behaviourists ignored or even denied – the bit that goes on in the brain, the bit we can't see – mental activity.

DECLARATIVE AND NON-DECLARATIVE MEMORY

Psychologists and neuroscientists distinguish between two types of memory – declarative and non-declarative memory.

Declarative memory is conscious and works with the type of knowledge that we learn, such as that the sky is blue or the supermarket shuts at 8pm. It is specific, relating to information that is either true or false.

Non-declarative memory works with skills and knowledge that we don't have to consciously think about, such as riding a bicycle or any conditioned responses. It also relates to habits and sensitization. It is expressed through behaviour and performance.

Often, declarative memory is said to be concerned with 'knowing what' and non-declarative with 'knowing how' (in the words of the English philosopher Gilbert Ryle).

There is not space to deal with all of the concerns of cognitive neuroscience, so we will focus on perhaps the two most important: memory, which is the basis of learning and personality, and our sense of identity, forged from personality, consciousness and the belief in free will.

Ideas of memory

Memory is an important mental function. It is vital to learning and social functioning. People with memory impairment often struggle to cope with everyday life.

THREE STORES FOR MEMORY

Clearly, not everything we see, hear, taste, smell or encounter is remembered. There is a process of filtering or selection going on. Psychologists divide memory into three types: sensory memory, short-term (or working) memory and long-term memory. Sensory memory is very transitory, held for a second or two. If you look across the room, everything in the scene is momentarily available to sensory memory. But you won't remember it a few minutes later. Anything that might be useful is moved to short-term memory (also called working memory). So if you saw someone you knew across the room, or perhaps saw smoke seeping under the door, this would be selected for retention and perhaps further processing.

We can use short-term memory to store nine or ten items for recall after a few minutes. For example, you could recall a phone number or items on a shopping list, but usually these will be lost from memory after a short time. More important information, which we actually want to learn, can be moved to long-term memory. As far as we know, this is unlimited in capacity and endurance. Something learned at age five can be recalled at age 95, and we can keep adding to our store of memories and knowledge throughout even the longest of lives. We are more likely to move things into long-term storage if we rehearse them (revisiting and reinforcing the memory).

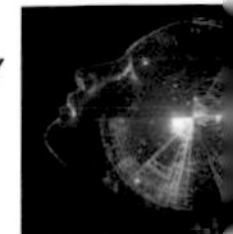

There are two aspects to consider in investigating memory: first, which parts of the brain are involved in forming, storing and recalling memories and, second, exactly how at a cellular level memories are formed, stored and recalled. The first is a problem in cognitive neuroscience; the second is a problem in molecular neurobiology. Considerable progress has been made on the former since the 1950s, but there is still a long way to go in understanding the latter issue.

Starting with cells

The earliest model of memory is that sensory stimuli enter the first cell at the front of the brain, are processed in the second cell and stored in the third cell at the back of the brain. The division of memory storage into three stages, corresponding with three locations, is echoed in modern psychological theories about separate memory stores (see box opposite).

The modern model of memory, too, identifies three processes involved in forming and using memories: encoding, storage and retrieval. Encoding covers the first stage – sensory input to the brain is interpreted and creates a memory trace. There are two sub-stages here: the acquisition of information from the senses and its consolidation. Storage is the passing of the encoded information to the part of the brain where it is kept. Retrieval is recalling the memory when it is needed.

Learning by association

Memory is a prerequisite for learning. The essential component of learning and comprehension is forming associations between different items of information or sensory inputs.

Aristotle believed we make mental links or 'associations' between sensations and events that are related in some way: we associate ideas that appear close together in time or space, or are similar, or often occur together, or even clear contrasts (such as hot and cold). Associations then form the basis of knowledge. We create a mental construct from elements of something we experience – so the physical appearance, smell and taste of, say, an orange, are all experienced together and associated, giving the experience and idea of an orange.

The idea was further developed in the 18th century by David Hartley. He

David Hartley considered vibrations of the nerves to be the origins of sensations and ideas.

PAVLOV'S DROOLING DOGS

Around 1903, the Russian physiologist Ivan Pavlov (1849–1936) was investigating the reflex that leads dogs to salivate when they smell or taste meat. He found that if he played a sound to the dogs before feeding them, they soon learned to associate the sound with the food and would then salivate on hearing the sound, even if he did not present any food. This is now known as classical conditioning.

'Give me a dozen healthy infants, well-formed, and my own specified world to bring them up in and I'll guarantee to take any one at random and train him to become any type of specialist I might select – doctor, lawyer, artist, merchant-chief and, yes, even beggar-man and thief, regardless of his talents, penchants, tendencies, abilities, vocations and the race of his ancestors.'

John B. Watson, 1913

focused on the grouping of ideas or impressions to create the package that represents an idea or experience. He tried to explain the transition from sensory perception to idea, proposing that sense perceptions produce vibrations in the nerves which travel to the brain, causing the brain to produce sensations. After an immediate sensation has passed, Hartley believed that echoes of the vibrations, which he called 'vibratiuncles', remain in the brain; this is the form taken by ideas. Simple ideas can group together into ever more complex ideas. Sensations experienced together become associated with one another and linked in the brain, so that one can call up the other in memory or when interpreting sensory input.

In the early 19th century, British philosopher James Mill (1773–1836) made association the foundation of everything that the mind can do. He believed associations explain the 'physics' of the mind in the same way that Newtonian physics explains the natural universe.

The domain of psychology

The early 20th century saw a great deal of work

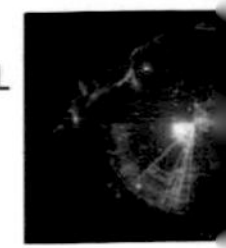

on memory and learning carried out by psychologists. Much of it was conducted by the behaviourists who maintained that only physical behaviours are susceptible to examination and, as mental states cannot be observed directly, they must be ignored – or might even not exist. A lot of their work was with animals. The starting point was Ivan Pavlov's work with classical conditioning in dogs (see box opposite). The pioneering behaviourist John B. Watson, greatly influenced by Pavlov, believed that classical conditioning can account for all types of learning in humans as well as animals, even including language. The belief that all behaviour is, in effect, programming, led to the conclusion that all outcomes can be manipulated. Taken to its logical conclusion, it would mean free will does not exist and that people can be moulded or engineered by controlling what happens to them and what they are exposed to in early life. This is a troubling finding, but one to which neuroscience would return.

ENGRAMS

The term 'engram' was coined by German zoologist Richard Semon in the early 20th century. He meant by it a memory trace that is indelibly encoded in the nerve cells and can be reactivated if one element of the original complex of stimuli is encountered again – so we can recall a scene or event from a single part (such as a smell or sight). Unfortunately, he spoiled a good idea by believing that these traces or changes in the brain could be inherited, so that memory units could pass from one generation to the next.

Making it physical

While philosophers and psychologists have dealt with theories and ideas about memory and learning, it has been the task of neuroscience to try to discover exactly what happens in the nervous system when we make and retrieve memories. It is a challenging task, and one that is far from complete.

Locating memory

The first experimental work trying to locate memory was carried out by the American neuropsychologist Karl Lashley (1890–1958). He experimented with rats, excising one part of the cortex after another and recording the results. He trained his rats to find their way through mazes, either before or after mutilation of the cortex, and then sought to find a localized trace of the memory of the maze in the brain (called an engram – see box above). His search was unsuccessful; he found simply that the more of the cortex was destroyed, the more the rat's abilities and memories were impaired. This he termed the law of mass action. He proposed instead, in 1929, that a memory is not stored in a single location, but is distributed across the surface of the brain.

Canadian psychologist Donald Hebb (1904–85) was the first to attempt a microbiological explanation for Lashley's 'mass action' findings. He related his explanation to the ancient concept of association: 'The general idea is an old one, that any two cells or systems of cells that are repeatedly active at the same time will tend

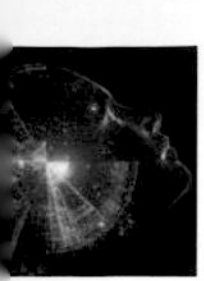

Rodents are useful but often doomed model organisms in neuroscience research projects.

to become "associated", so that activity in one facilitates activity in the other.'

Hebb found that a new perception or a shift of attention fires a package of neurons together, which he called a 'cell assembly'. He gave as an example a child hearing footsteps and then seeing a parent approach. The footsteps trigger a cell assembly forming a package of perception. The same stimulus – hearing footsteps – will excite the same assembly next time. Seeing the parent triggers another cell assembly. If the parent appears soon after the footsteps, the two assemblies can be linked, forming what he called a 'phase sequence'. The next time the first cell assembly is triggered by hearing footsteps, the child anticipates the arrival of the parent.

This is a clear, biological explanation of association and of learning. It is most commonly expressed in the form of reinforcing neural pathways: the more two (or more) neurons are linked by consecutive firing, the stronger the link between them grows and the more likely firing one is to lead to the firing of the other. Because the neurons involved in an assembly or phase sequence can be distributed over the cortex, the memory can sustain limited damage to the cortex, explaining Lashley's results with his rats.

The case of H.M.

In 1953, a patient called Henry Molaison (known as H.M.) underwent surgery for intractable epilepsy. His surgeon removed the parts of the brain determined to be causing the epilepsy, the medial temporal lobes. After the operation, H.M. suffered severe memory impairment. His condition was reported by Brenda Miller in 1957, who had previous experience of a similar patient after removal of the hippocampus. H.M. could not form new memories or learn new vocabulary, nor remember things that he had done from one day to the next. He could not recall memories formed during the three years prior to his operation. Yet he showed no intellectual loss or degraded perceptions. The process of consolidation – rooting a memory firmly – involves moving it from one part of the brain to another. The conclusion drawn from H.M's case was that the medial aspect of the temporal lobe is vital to memory. His unfortunate experience kick-started modern cognitive and neurological work on memory.

In fact, something even more complex emerged from the case of H.M. Although his declarative memory was severely

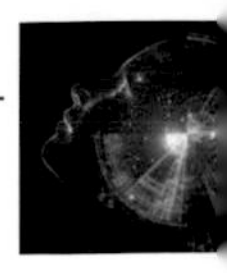

> *'If the inputs to a system cause the same pattern of activity to occur repeatedly, the set of active elements constituting that pattern will become increasingly strongly interassociated. That is, each element will tend to turn on every other element and . . . to turn off the elements that do not form part of the pattern. To put it another way, the pattern as a whole will become "auto-associated". We may call a learned (auto-associated) pattern an engram.'*
>
> Donald Hebb, 1949

impaired, he was able to form non-declarative memories. He could acquire new motor skills, but was not able to say that he learned them – he knew *how* to do something, but did not know *that* he knew how to do it. This, then, suggested that non-declarative memories are not formed in the hippocampus. Further, he could maintain attention for a long time and retain information for a short period, suggesting that short-term or working memory is not located in the medial temporal lobe. As he could recall memories formed long before his surgery, it was clear that long-term memory is not located in the area that was removed. It was assumed that long-term storage happens in the neocortex. Finally, his unimpaired intellectual and perceptual function demonstrated that these do not rely on the medial temporal lobe.

H.M.'s case led to the biological distinction between declarative and non-declarative memory. It became clear that 'non-declarative memory' is not really one type of memory, but an umbrella term for everything not susceptible to conscious and deliberate recall. It was declarative memory specifically that H.M. had lost. Non-declarative memory includes all the accumulated habits and preferences that make us who we are as individuals – H.M.'s character therefore remained intact.

That H.M could remember more distant personal memories – those formed three or more years before his surgery – suggests that the role of the medial temporal lobes in retaining a memory decreases over time. Research with other

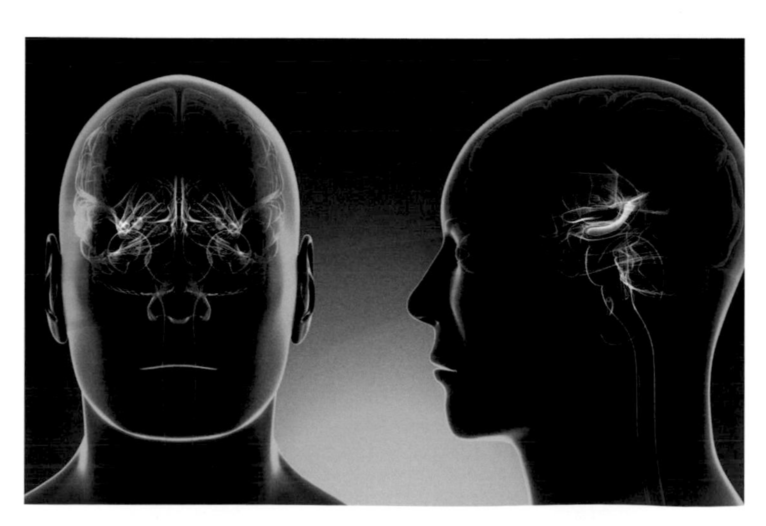

The hippocampus is deep within the brain, in the medial temporal lobe on each side.

patients and animals has borne out this conclusion. A study with mice reported in 2005 found that activity in the hippocampus gradually decreased after learning but activity in several regions of the cortex increased, suggesting the burden of processing and storing the new learning was shifting from the hippocampus to elsewhere.

Current theory suggests that some memories are stored in the area originally responsible for receiving and processing the sensory information, so visual memories should be stored in the area of the brain responsible for processing visual information. This seems to be borne out by the experience of a painter, reported by psychologist Oliver Sacks in 1995. An accident rendered him colour-blind, probably by damaging the part of the brain involved in colour perception, and he was not only unable to see things in colour but was also unable to remember or visualize in colour. A German study using EEG reported in 2016 that the same areas of the brain are activated when encoding a memory and then when later retrieving it, also suggesting that some memories are stored where the perception is first formed.

Tests on other subjects with different specific forms of amnesia and different lesions show certain categories of information are lost, suggesting that the way information is stored in the brain (and the location of storage) depends on many aspects of the information, such as whether objects are usually defined by their use or their characteristics.

The conclusion of more than 50 years of study, starting with H.M., is that the medial temporal lobe, and specifically the hippocampus and the region immediately around it, is involved in processing and consolidating information in working memory for long-term storage as declarative memory. Long-term storage is distributed around the neocortex, with elements of a composite memory being stored (and later retrieved) in the areas originally involved in their perception. The role of the hippocampus in consolidating memories in their appropriate locations can take years – hence the gap in H.M.'s memory of three years prior to his surgery.

Neural pathways

As we have seen, Hebb suggested it is the connections between neurons in the brain which are crucial to learning and memory. Ramón y Cajal had also concluded that, in the adult brain, neural cells have lost the ability to divide and reproduce, so the brain's plasticity must lie in growing branches to form and strengthen networks between the cells.

> *'Let us assume that the persistence or repetition of a reverberatory activity (or "trace") tends to induce lasting cellular changes that add to its stability. . . . When an axon of cell A is near enough to excite a cell B and repeatedly or persistently takes part in firing it, some growth process or metabolic change takes place in one or both cells such that A's efficiency, as one of the cells firing B, is increased.'*
>
> Donald Hebb, 1949

The colourful sea-slug Aplysia found a career as a model animal for neuroscientists.

The question remained of how this works at the molecular and cellular level. How are changes to the neurons effected? Are the same types of changes involved in short-term and long-term memory, in declarative and non-declarative types of memory? A model organism was needed for the research – the lucky creature chosen was Aplysia.

Slugs with memories

Aplysia is a type of large sea slug. It has a reflex action, withdrawing its gill and syphon mechanism, in response to a potentially threatening stimulus. It's particularly useful as a test organism as it has a small number of large, easily visible neurons, and individual behaviours can be linked to small groups of around 100 neurons. It's possible to create a memory in Aplysia, then investigate the neurons to find which have changed and locate the memory that has been formed.

Working with Aplysia in the 1960s and 1970s, Austrian-born neuroscientist Eric Kandel demonstrated that learning is not accomplished by building new connections between neurons but by strengthening the pathways that already exist. This is effected by reinforcing the synaptic connections between neurons. Kandel studied three types of learning response in Aplysia:

- Habituation – the animal becomes used to a stimulus and its response decreases
- Dehabituation – a new stimulus causes the response to occur again
- Sensitization – the animal becomes sensitized to a stimulus, so the response strengthens (becomes more pronounced).

How it works

When Aplysia receives a sensitizing stimulus (such as a light electric shock) the sensory nerve stimulated releases the neurotransmitter serotonin. This modulates the strength of the connection between the sensory neuron and the motor neuron. Before the animal has learned about the stimulus, an action potential in the sensory neuron produces a small potential in the motor neuron. After the animal has been sensitized, though, an action potential in the sensory neuron produces a larger potential in the motor neuron. This increases the likelihood that each connected motor neuron will be activated so will produce a greater response (larger contraction of the muscle). Soon, the same triggering of the sensory nerve produces a greater result – the animal has become

sensitized. This happens because the connection between the sensory and motor neurons has been strengthened.

Forming a short-term memory involves modulating the channels in a neuron's membrane through which chemicals pass. The biochemical changes that result are short-term. The same type of biochemical mechanism is involved in all short-term memory, including our own memory for things we want to recall for a few minutes. If you memorize a phone number just long enough to use it, for instance, you store it in short-term memory, using the same biochemical systems as Aplysia uses to 'remember' it has just been poked, even though in the case of Aplysia it is not conscious memory but reflex.

The results of studies with Aplysia and other simple organisms revealed in the 1970s that non-declarative memories do not need any special neurons or organs within the brain, but are stored in the same neurons that form part of the reflex pathway. Different types of learning/memory can be stored, and learning/memory can be distributed along the pathway.

'A prerequisite for studying behavioral modification is the analysis of the wiring diagram underlying the behavior. We have, indeed, found that once the wiring diagram of the behavior is known, the analysis of its modification becomes greatly simplified.'

Eric Kandel, 1970

Exploring long-term memory

Long-term memory is a very different mechanism. Instead of the transient chemical changes within a neuron that produce short-term memory, long-term memory involves changes in the structure of neurons. This is called long-term potentiation (LTP).

LTP was first observed in the hippocampus of the rabbit in 1966. In Oslo, Norway, Terje Lømo discovered that if he delivered a series of high frequency stimuli to a pre-synaptic neuron and then a single pulse stimulus, the effect on the post-synaptic neuron lasted much longer than if he delivered the single pulse alone. The post-synaptic neuron had become potentiated by the rapid series of stimuli.

Exactly how LTP works is still not known. Neurons are able to grow (and lose) new processes called dendritic spines which are thought to be involved in memory storage and making connections between neurons. Each dendrite can have thousands of spines. Kandel's findings with Aplysia suggested that neuroplasticity does not extend to making entirely new connections but strengthens or reduces existing links between neurons. He found that the basic neural 'wiring' is already in place and is inherited; the impact of experience is to forge preferred pathways from the basic structure (or to allow pathways to erode).

Super-smart mice

Forming dendritic spines involves using proteins and altering gene expression. At least 25 genes are involved, and a corresponding number of proteins, and neuroscientists are still exploring this area.

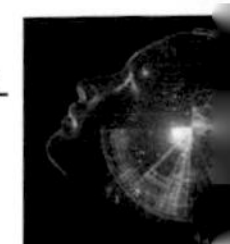

The basic mechanism came to light in 1996 with the work of the Chinese brain researcher Joe Tsien, working at Princeton University. Tsien used genetic engineering techniques to create a transgenic mouse that had extra NMDA (N-methyl-D-aspartate) receptors and, as expected, found the mouse smarter than a normal mouse. Tsien had previously found that limiting the expression of the gene that controls NMDA production resulted in stupider mice.

The super-smart mice, dubbed Doogie, could learn faster and remember information longer than the control mice. This was the first molecular confirmation of Hebb's theory. It holds out hope, too, of treatments for memory impairments in human patients. But in 2001 researchers found that the Doogie mice are more susceptible to chronic pain. Any treatment for memory impairment that targets NMDA will have to take account of this vulnerability, and any attempt to limit NMDA to

MAKING AND BREAKING MEMORIES

In 2014, researchers in the University of California working with rats managed to both remove memories and strengthen memories. They used a technique called optogenetics, which adds a light-sensitive gene to neurons and then activates the neuron by shining a bright light at it. The experiment provided the first proof of the theory that long-term potentiation is at the base of memory. The researchers succeeded in strengthening the rat's memory, reinforcing the connection between neurons, and eradicating the memory by weakening the connection. They could even reinstate the memory later by strengthening the connection again. It's a strategy that might one day be used to help people suffering from post-traumatic stress disorder (by removing or reducing memories) or those with conditions involving memory impairment, such as Alzheimer's.

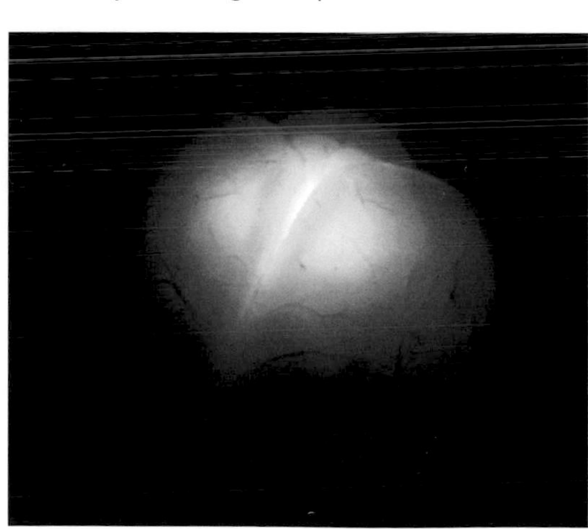

A miniature LED embedded in a mouse's brain provides researchers with a way of triggering neurons less intrusively than using a conventional probe.

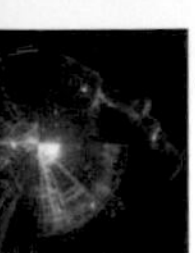

control pain will have to take account of possible interactions with memory. Neruroscience is riddled with such complications.

Back to the hippocampus

The case of H.M. revealed that the hippocampus is involved in forming, though not storing, long-term memories. It is also essential in spatial reasoning and navigation. 'Place cells' in the hippocampus have been found to help mice identify locations and know where they are. These were first identified in 1971. When the animal is put in a new environment, it has to make a new 'map' in the hippocampus to understand and remember locations. An important study involving London taxi drivers demonstrated that the hippocampus is important in spatial memory and navigation in humans, too.

'I had that neuron in the back of my cab . . . '

A study in 2000 found that London taxi drivers have a larger hippocampus than members of a control group. Further studies involving trainee taxi drivers showed that as taxi drivers successfully acquired 'the Knowledge' (detailed memory of London's streets), the volume of grey matter in the mid-posterior hippocampus grew and the volume in the anterior hippocampus diminished. Comparison with bus drivers (who are subject to similar daily routines and environments but follow pre-determined routes) suggests that the change is directly related to learning and storing complex spatial information. The hippocampus is one of the few areas of the brain able to grow not just new neural connections, but entirely new neurons in adulthood. Taxi

London taxi drivers have to learn their way around the city's streets – a formidable feat of memory.

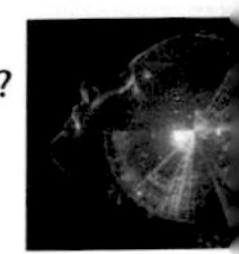

Our experiences combine to help build our character. Overall, a pattern of happy experiences and feeling secure helps build a confident individual – trauma and misfortune affect character adversely.

drivers did less well on other visual learning and memory tasks, suggesting that there is a cost to acquiring their specialist knowledge.

It was found that the hippocampus is also important in recall, at least for episodic memory. When a person remembers a life event, such as a family outing, the hippocampus brings together many aspects of a scene, including sounds, sights and smells, stored in different parts of the cortex.

Who do you think you are?

Most people have a natural sense that their identity is produced by their mind and while it might be located in their brain it is not exactly the same as the brain. Neuroscience suggests that this might not be the case, but that the brain and mind are one and our personalities are entirely forged by the connections between neurons in our brains.

One of the things that memory does is construct personality. We are all of us built from our previous experiences and their conscious and subconscious effects, from what we have learned and the consequences of past actions. Personality informs our choices and actions – and yet we like to feel we have free choice in what we do. The extent to which character is determined by neural physiology is contentious. Unless we

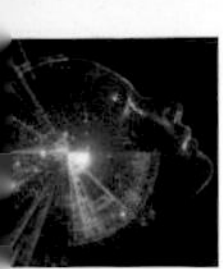

Phineas Gage, photographed following his terrible accident in 1848.

accept some kind of metaphysical soul that influences our thoughts and actions, we must decide that we are entirely determined by the material structure of our brains and the pattern of neural connections that has been built up through experience and laid down by genetics.

The phrenologists Gall and Spurzheim were convinced that the personality is determined by the physical size of different organs linked with character traits or qualities such as benevolence or curiosity. Around the same time as phrenology was enjoying success, an accident on the American railroads produced a new insight into the parts of the brain involved with constructing personality.

A blast in the past

One of the most famous cases in neurology is that of Phineas Gage, an American railroad worker. Gage was in charge of tamping down explosives ready for blasting rock apart to lay railroads. In an unlucky accident in 1848, a tamping iron (like a crowbar) was blasted through his head, entering below his left eye and leaving through his skull, taking part of his brain and skull with it. Against all the odds, he survived, but not without some adverse effects. As well as scarring and the loss of an eye, Gage suffered mental changes. He went from being a gregarious and cheerful person to one described by his physician, John Harlow, as 'fitful, irreverent, and grossly profane, showing little deference for his fellows' and 'capricious and vacillating'. Although in 1850 the professor of surgery at Harvard University, Henry Bigelow, reported that Gage had 'quite recovered in faculties of body and mind', Harlow wrote in 1868 that his personality had changed so radically that his friends and acquaintances said he was 'no longer Gage'. Gage died after a series of

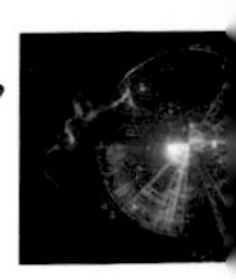

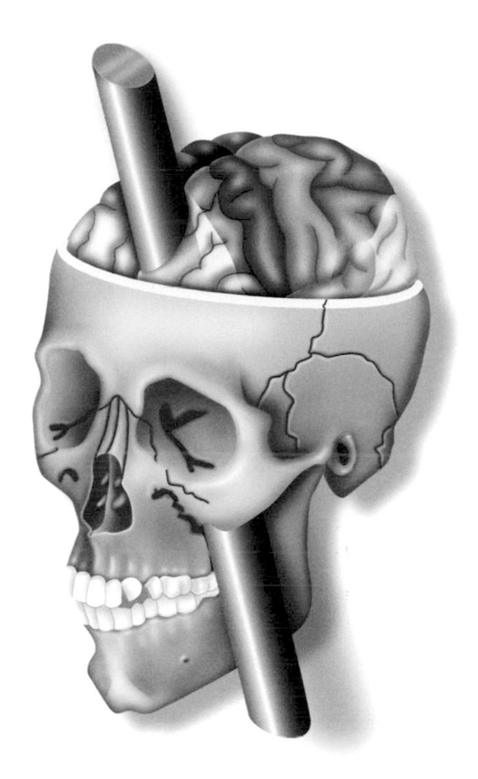

The path of the tamping iron through Gage's head, damaging the frontal lobe.

epileptic seizures in 1860. It's hard to judge now how much his character had changed; he certainly lost his job on the railway, but was able to work as a stagecoach driver for a considerable period. Modern researchers suggest the routine and predictability of his work helped Gage to cope.

Gage's accident provided some of the first evidence that the frontal cortex is involved in personality. Gage was used as a supporting case by the Scots neurologist David Ferrier in 1878 when reporting his work with primates. Ferrier had found that damage to the frontal cortex did not affect the animals' physical abilities, but produced 'a very decided alteration in the animal's character and behaviour'.

Researchers continue to work with Gage's skull. In 2012, Jack Van Horn working at the University of California in Los Angeles produced a digital model of the rod's path through the skull which suggested that up to 4 per cent of the cerebral cortex and more than 10 per cent of the total white matter could have been destroyed. In addition, he lost connections between the left frontal cortex and other areas of the frontal cortices, the limbic structures. The loss of these connections within the brain might have been more important than the damage to the left frontal cortex itself in explaining his changed behaviour.

Phrenology revisted?

Before the development of brain-imaging technology, it was not possible to work on the correlation between brain structure and personality in any way other than studying

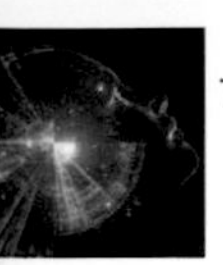

the effects of lesions or damage. But now, using EEG and fMRI, it's possible to see which parts of the brain fire when a person acts or responds in a certain way. For example, researchers can show someone a distressing image and observe which areas of the brain are immediately active. These parts might be correlated with emotional distress – or they might not. Unfortunately, the brain is not labelled, and it's equally possible the activity corresponds to a desire to get away from the stimulus, or perhaps something else entirely. It takes a lot of tests and a lot of cross-referencing to come up with some designations that neuroscientists agree on, and many fMRI studies use small sample sizes.

In the late 20th and early 21st centuries, a large number of fMRI studies claimed to find areas of the brain that were positively correlated with emotional activity, such as empathy or social anxiety, or with personality traits. It is not certain, though, that more brain activity means a stronger tendency or response. In some cases, the opposite is true, with experts in a task using less brain power to accomplish it than novices, who have to struggle and concentrate. Criticism in 2008 cast doubt on the validity of many fMRI studies into personality, emotion and social cognition.

How do you feel?

Making sense of the areas shown up by fMRI scans is vital if they are to produce meaningful information. A study in Pittsburgh, USA, in 2013 scanned the brains of professional actors rehearsing a series of emotions and then fed all the images into a computer so that it could 'learn' the brain patterns associated with them. To check that the patterns were not different because the actors were simulating the emotions, they compared them with genuine experiences of these emotions. The computer-learning system could then identify the same emotions in a new subject with reasonable accuracy, suggesting that there are generalizable patterns to the brain activity of certain emotions. Being able to read the emotions of a research subject avoids the unreliability of self-reporting. The study revealed that areas widely distributed around the brain are involved with emotional responses.

Scanning techniques such as fMRI are a long way from being able to give the kind of character breakdown that the phrenologists laid claim to. It reveals current activity, not patterns of behaviour or thought. We might be able to observe the brain activation patterns of a person feeling kindly towards someone at one moment, but cannot see whether there is a predisposition towards kindness. For now, reading personality is beyond the horizon.

Making choices

Fundamental to our sense of self is the belief that we are in control of our thoughts and actions. It can feel as though the concept of free will is compromised if we allow too much influence to either genetics or environmental impacts (nature or nurture) on the development of character. This might seem particularly true if we consider the neurological encoding of these influences. If the mass of neural connections in our brains, formed through a mix of DNA and memories,

CAN DEAD SALMON READ HUMAN EMOTIONS?

The interpretation of fMRI scans involves using software which has to be set with tolerances that balance eradicating 'noise' (meaningless signals) with missing genuine data. A neuroscience researcher, Craig Bennett, showed the dangers of getting this wrong in 2009 when he put a dead salmon into an fMRI scanner, then exposed it to pictures of humans displaying different emotions and asked it to identify the emotions. The raw data from the scanner showed orange pixels, identifying activity, in areas of the salmon's brain suggesting it was indeed thinking about or responding to the pictures. Bennett's conclusion was not that dead salmon are able to detect emotions, but that careless use of fMRI data may deliver unreliable results.

determines our thoughts and actions, how much free will do we really have? This question has been tackled by some studies in decision-making.

The Libet experiment

In 1983, the American neurophysiologist Benjamin Libet carried out an experiment to determine the time lag between a person's brain showing a decision has been made and the person being aware of making the decision. A subject had to choose a random moment at which to flick their wrist, noting the position of a dot on a clock when they made the choice. The subject was connected to an EEG that measured brain activity throughout.

The experiment made use of the build-up of electrical signals called the readiness potential before a physical action takes place. In 1965, Libet found that the change in readiness potential typically came about half a second before the subject was aware of having formed the intention to move. Libet's conclusion was that the unconscious decision is made before we become aware of the decision – so we believe ourselves to be making a conscious decision, but are

The sense of free will gives meaning to our lives. If instead we believe all our actions are predestined by our biology (or by a divinity) we need to look elsewhere for a sense of purpose, value and agency.

actually only becoming aware of a decision that has already been made unconsciously.

Later studies generally support the finding that awareness comes some time – measurable in seconds or fractions of a second – after brain activity, which indicates some kind of decisive move towards action has already started. A revised version of Libet's experiment carried out in 2008 removed the need for the subject to give their account of when they made (or noticed) their intention to move, reading this information, too, directly from the brain. The result was that sometimes the participant didn't become aware of the decision until after they had started to move – suggesting noticing the movement was interpreted as making the choice. In 2011 the American neurologist Itzhak Fried investigated decision-making at the level of tracking individual neurons as they fired. He found a lag of up to two seconds between a neuron firing and the subject becoming aware of their decision.

The obvious conclusion, that free will is an illusion, may simply show that we attribute too much importance to consciousness in our everyday actions. This issue may be one of semantics as much as philosophy. What we mean by 'free will' and how we define its relationship to awareness may be as important as the experimental evidence itself. The Canadian philosopher Daniel Dennett has said that the kind of free will which these experiments would deny is a type that is not worth having.

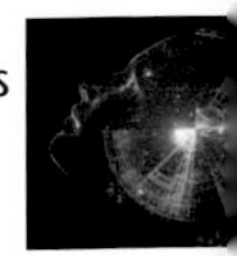

Libet pointed out that the conscious will could still veto the decision at the last moment and the half-second recorded as the build-up to consciousness of the choice might be only a stage of preparedness which could be abandoned.

Acting on choices

From the 1970s, neuroscientists have had techniques for studying the activity of single neurons. Beginning with work on monkeys, Edward Evarts and Vernon Mountcastle were able to show correlations between cognitive processes such as perception and decision-making and the firing patterns of individual neurons. It became possible to trace the exact neural pathway from stimulus, through processing, to behaviour.

Single-neuron techniques can now be applied to the human brain. They have applications in mapping brain activities, in neural pathways typical of some neurological conditions (such as

CAN WE READ YOUR THOUGHTS?

Mind-reading has long been a popular topic in science fiction and fMRI looks as though it could be the way to make it happen in reality. During the second decade of the 21st century, various studies tracking the neural activity of subjects' brains have been able to reconstruct words from auditory information in the brain, to approximate visual signals and to control objects or elements on a computer by tapping into firing neurons. It looks likely that fMRI could provide a lie-detecting mechanism, but it's not there yet. There are many ethical implications to consider before using technology to read someone's brain, but there are also clear clinical applications, including helping patients who are unable to communicate, such as those in a permanent vegetative state.

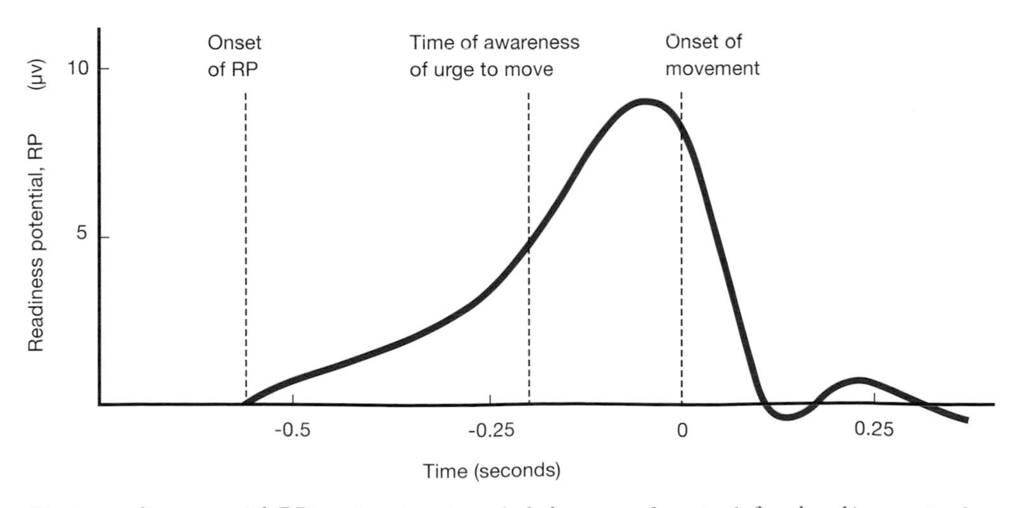

Plotting readiness potential (RP) against time shows the body prepares for action before the subject consciously decides to act.

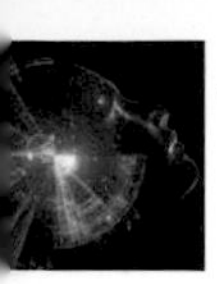

A Japanese study in 2015 found a computer could predict from brain scans which choice people playing the game rock, paper, scissors would make before they moved.

Parkinson's disease) and, most dramatically, in brain–machine interfaces (BMIs). These use electrodes in the brain to pick up the electrical signals associated with the firing of a neuron before passing them to a computer. Ultimately, the aim is to take the intention to move and pass it to a prosthetic device, enabling patients with paralysis or loss of limb to communicate or move. The technology is not yet good enough but the concept is in place – choices made in the brain (whether consciously or not) can be interpreted and passed to a computerized device to implement.

Is there anybody there?

Descartes famously wrote: 'I think, therefore I am.' He took for granted the existence of an 'I', which we would probably call consciousness. But consciousness itself is difficult to define.

In 1995, Australian philosopher and cognitive scientist David Chalmers defined what he called the 'hard problem of consciousness'. The easier problems of consciousness he identified as those we can plausibly approach using computational or neural mechanisms. Among these he listed features such as the focus of attention, the difference between sleep and wakefulness and the deliberate control of behaviour. These are not yet fully explained, but neuroscientists and cognitive scientists have ways of approaching them empirically.

The 'hard' problem of consciousness, though, resists these and other approaches. Trying to define consciousness, he said: 'an organism is conscious if there is something

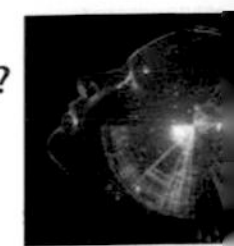

it is like to be that organism, and a mental state is conscious if there is something it is like to be in that state.' With the very nature of consciousness, we are at an impasse. Consciousness is the bit that produces experience from the raw material the mind has to work with. A dog or a cat or a worm might see what we see but, we suspect, does not experience the phenomena in the same way as we do. Even individual humans might not experience the same phenomena in the same way. In Chalmers' words:

'This . . . is the key question in the problem of consciousness. Why doesn't all this information-processing go on "in the dark", free of any inner feel? Why is it that when electromagnetic waveforms impinge on a retina and are discriminated and categorized by a visual system, this discrimination and categorization is experienced as a sensation of vivid red? We know that conscious experience does arise when these functions are performed, but the very fact that it arises is the central mystery. There is an explanatory gap . . . between the functions and experience, and we need an explanatory bridge to cross it.'

There's no such thing as a soul?

The hard problem of consciousness seems to bring neuroscience up short: everything else appears to be susceptible to physicalist explanation of one sort or another, even if it depends for its detail on things not yet fully explained, but consciousness cannot be explained in this way. We are back with Descartes, looking at the unbridgeable gap between body and soul, recast in 21st-century garb. The difference is that we don't necessarily share Descartes' confidence that there is a spirit.

The question we are left with is not addressed by neuroscience: whether the mind, all cognitive activity, is entirely produced by the action of neurons or whether there is something else – akin to a soul, perhaps – separate from the physical and chemical activities of the brain. There are neuroscientists who believe there is a soul, and neuroscientists who believe there is not.

Playing a complex piece of music does not require conscious awareness of every note – some freely chosen acts do not rely on continuous conscious engagement.

INTO THE FUTURE

Developments in neuroscience have had an impact on disciplines as diverse as philosophy, computer science, law and linguistics and will do so increasingly.

Intelligence – artificial or otherwise

At present, artificial intelligence is very unintelligent. An approach using artificial neural networks aims to change that. It tries to mimic the way the human brain reinforces or weakens neural connections in the process of learning. An artificial neural network has a collection of neural units and 'learns' from exposure to many examples or situations, forging its own connections as it goes. The most advanced artificial neural networks currently have only a few million neural units and millions of connections – about the cognitive potential of a not-very-bright worm. By contrast, the human brain has around 84 billion neurons, some with thousands of connections each. There's a long way to go. Ironically, neuroscientists use AI systems to identify patterns of neural activity which correspond to particular stimuli or responses.

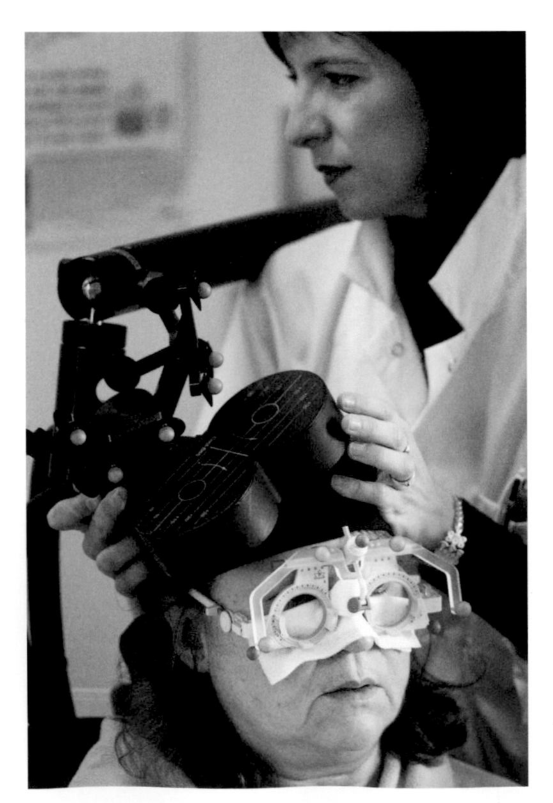

Transcranial magnetic stimulation uses magnetic fields to prompt electrical currents in the brain and has recently started being used to treat depression and eating disorders.

Brain control for good and bad

As in many areas of scientific development, there are threats as well as potential benefits in what neuroscience promises to bring us in the coming years. Measuring or harnessing brain activity could be used to help people with disabilities to control their surroundings or to speak, for example. But examining brain activity could be used to tell whether someone is lying, to read or perhaps even change their thoughts. In India, in 2008, a woman was convicted of the murder of her ex-fiancé after EEG evidence showed she was familiar with details of the killing. Knowing how to trigger sensations could be used to create either pleasure or pain with no physical correlate. There is a need for ethical philosophers and legal experts to work alongside neuroscientists.

But really, where are you?

Neuroscience is concerned with the physical and chemical goings-on of the nervous system. It can't, at present, approach such notions as why one person is more compassionate or musical than another, how a creative idea or inspiration springs into the mind, why we experience things in the way we do, or what makes us choose one long-term course of action over another. The ghost in the machine – if there is one – remains as elusive as ever. And if there isn't one, perhaps that is even more miraculous.

We still have no idea where the impulse to make art comes from, or how creativity and inspiration arise and are translated into physical movement.

PASSTHOUGHTS

If you have difficulty remembering passwords, you might welcome the possibility of passthoughts. An EEG scan that identifies your brainwaves uniquely could one day replace some other methods of security clearance. Work on passthought technology in California in 2013 achieved greater than 99 per cent accuracy.

'How can a three-pound mass of jelly that you can hold in your palm imagine angels, contemplate the meaning of infinity, and even question its own place in the cosmos?'

V.S. Ramachandran, neuroscientist, 2011

Index

Picture credits

Alamy Stock Photo: 23 (Mary Evans Picture Library), 96 (Granger Historical Picture Archive), 117, 138–9, 174(t)

Bridgeman Images: 48, 116, 154–5

Diomedia: 103 (Natural History Museum, London, UK), 118

Getty Images: 14 (UIG), 21 (Bettmann), 24 (Science & Society Picture Library/Science Museum Pictorial), 30, 64, 77, 78, 90–91 (AFP), 95(r), 106 (The Asahi Shimbun), 153 (UIG), 162, 165 (Bettmann), 166, 170 (SSPL/Science Museum), 172 (ullstein bild), 184 (UIG), 195 (UIG)

123RF: 25 (kmiragaya)

Laboratory of Neuroimaging and Martinos Center for Biomedical Imaging, Consortium of the Human Connectome Project: 111

Science & Society Picture Library: 42–3 (Science Museum Pictorial), 191

Science Photo Library: 10, 46, 95(l), 97, 122, 146, 164 (National Library of Medicine), 177, 194 (Science Source), 202

Shutterstock: 6, 8–9, 12, 17, 22, 32, 34, 44, 61, 62, 65, 70, 99, 101, 105, 107(b), 110, 112–13, 120, 121, 124(b), 125, 128, 129, 135(x2), 137, 140, 141, 143, 144, 145, 147, 151, 167, 168–9, 174(b), 175, 176, 178–9, 180, 181, 182, 186, 187, 189, 192, 193, 197, 198, 200, 201, 203

Wellcome Library, London: 13, 16(x2), 26, 28–9, 35, 37, 38, 39, 40, 49, 50, 51, 52, 53, 54, 55, 56, 57, 58, 60, 68–9, 72, 73, 76, 79, 81, 82, 83, 85, 86, 87, 92, 93, 94, 108, 115, 119, 124(t), 126, 127, 130, 133, 148, 149(t), 156, 157, 158, 159, 160, 161, 171, 183

Artwork by David Woodroffe: 7, 18, 107(t), 199